Developments in Surface Contamination and Cleaning

Developments in Surface Contamination and Cleaning
Volume Six

Methods of Cleaning and Cleanliness Verification

Edited by
Rajiv Kohli and K.L. Mittal

AMSTERDAM • BOSTON • HEIDELBERG • LONDON
NEW YORK • OXFORD • PARIS • SAN DIEGO
SAN FRANCISCO • SINGAPORE • SYDNEY • TOKYO

William Andrew is an Imprint of Elsevier

William Andrew is an imprint of Elsevier
225 Wyman Street, Waltham, 02451, USA
The Boulevard, Langford Lane, Kidlington, Oxford OX5 1GB, UK

First edition 2013

Copyright © 2013 Elsevier Inc. All rights reserved. With the exception of Chapter 3, Copyright owned by James T. Snow, Masanobu Sato, and Takayoshi Tanaka

No part of this publication may be reproduced or transmitted in any form or by any means, electronic or mechanical, including photocopying, recording, or any information storage and retrieval system, without permission in writing from the publisher. Details on how to seek permission, further information about the Publisher's permissions policies and arrangements with organizations such as the Copyright Clearance Center and the Copyright Licensing Agency, can be found at our website: www.elsevier.com/permissions.

This book and the individual contributions contained in it are protected under copyright by the Publisher (other than as may be noted herein).

Notices

Knowledge and best practice in this field are constantly changing. As new research and experience broaden our understanding, changes in research methods, professional practices, or medical treatment may become necessary.

Practitioners and researchers must always rely on their own experience and knowledge in evaluating and using any information, methods, compounds, or experiments described herein. In using such information or methods they should be mindful of their own safety and the safety of others, including parties for whom they have a professional responsibility.

To the fullest extent of the law, neither the Publisher nor the authors, contributors, or editors, assume any liability for any injury and/or damage to persons or property as a matter of products liability, negligence or otherwise, or from any use or operation of any methods, products, instructions, or ideas contained in the material herein.

Library of Congress Cataloguing-in-Publication Data
A catalogue record for this book is available from the Library of Congress

British Library Cataloguing in Publication Data
A catalogue record for this book is available from the British Library

ISBN: 978-1-4377-7879-3

For information on all Elsevier publications visit our website at www.store.elsevier.com

Contents

Preface — vii
About the Editors — xi
Contributors — xiii

1. **Removal of Surface Contaminants Using Ionic Liquids** — 1
   Rajiv Kohli

2. **Microemulsions for Cleaning Applications** — 65
   Lirio Quintero and Norman F. Carnahan

3. **Dual-Fluid Spray Cleaning Technique for Particle Removal** — 107
   James T. Snow, Masanobu Sato and Takayoshi Tanaka

4. **Microbial Cleaning for Removal of Surface Contamination** — 139
   Rajiv Kohli

5. **Cleanliness Verification on Large Surfaces** — 163
   Darren L. Williams and Trisha M. O'Bryon

Index — 183

Preface

The purpose of the book series *Developments in Surface Contamination and Cleaning* is to provide a continuous state-of-the-art critical look at the current knowledge of the behavior of both film-type and particulate surface contaminants. The first five volumes, published in 2008, 2010, 2011, 2012, and 2013, respectively, covered various topics dealing with the fundamental nature of contaminants, their measurement and characterization, and different techniques for their removal. This book is the sixth volume in the series.

The individual contributions in this book provide state-of-the-art reviews by subject matter experts on cleaning and cleaning verification.

Ionic liquids (ILs) and deep eutectic solvents (DES) are a new class of low melting point materials with unusual and unique properties that make them attractive for cleaning applications. In his first contribution, **Rajiv Kohli** discusses the characteristics of these solvents that include high solubility for a wide range of contaminants, thermal and chemical stability, low melting point (even below 273 K), very low volatility, and high conductivity. IL solvents do have some disadvantages including high cost, complex synthesis, need for purification for reuse, and high toxicity and nonbiodegradability of many formulations. Several DES formulations have been developed to overcome some of these limitations. Recently, cleaning applications have been proposed and have been successfully demonstrated, although many of these are largely still at the laboratory stage. The applications range from brush cleaning for removal of microcontaminants, semiconductor wafer and integrated circuit cleaning, precision cleaning of parts for aerospace applications, oxide-scale removal on metals, electropolishing of metals, microbial decontamination, cleaning of artworks, cleaning of wellbores in oil and gas recovery, soil decontamination, and consumer product applications.

Lirio Quintero and **Norman Carnahan** review the use of microemulsions for cleaning applications. The distinctive properties of microemulsions (high solubilization of oil, low interfacial tensions, and spontaneous formation) make these fluids very attractive for a variety of cleaning applications. Microemulsions have been used in industrial cleaning processes and household cleaning applications. Some of the applications include domestic and industrial laundry; cleanup of wastewater; cleaning of contaminated soil, fabrics and textiles, frescoes, paintings, monuments and building structures that have been polluted; and various applications in the oil and gas industry.

Cleaning processes will continuously be challenged because of the further scaling down of device structures and introduction of new materials and three-dimensional device features. Removal of these "killer" particle defects must be carried out without damage to these fragile features and with essentially zero material loss or roughening of exposed surfaces. The contribution by **James T. Snow, Masanobu Sato and Takayoshi Tanaka** discusses dual-fluid spray cleaning which offers high potential for removal of contaminant particles from various surfaces. The utilization of the forces created from the droplet impact phenomena from dual-fluid spraying has proven to be one of the more effective techniques for particle removal without damage in semiconductor wafer cleaning processing. The spray nozzle enables separate control of droplet diameter and velocity to provide optimized droplet energy. This has improved the cleaning efficiency and has reduced the potential for pattern damage caused by variations in droplet size and velocity.

The second contribution by **Rajiv Kohli** is an overview of microbial cleaning that takes advantage of naturally occurring microbes to remove a wide variety of contaminants from various surfaces. Microbial cleaning has been shown to be an effective alternative to conventional solvent cleaning for many applications. The method is based on the affinity of microbes for hydrocarbons which are digested, producing harmless carbon dioxide, water and soluble fatty acids. The microbes are nonpathogenic and are safe to handle and dispose. The process is environmentally friendly and is less expensive than solvent cleaning, but it is not applicable to high-precision cleaning applications. Typical applications include parts washing; oil and grease removal from concrete and other floor surfaces, and from drains and grease traps in manufacturing facilities, hospitals, restaurants, food processing facilities, and similar locations; cleaning of historical artworks and structures; cleaning and disinfection in health care facilities; wound debridement; controlling sulfate-reducing bacteria in oil fields; mercury bioremediation; and household and institutional cleaning applications.

Cleanliness verification is growing in its importance in many industries, such as aerospace, biomedical engineering, and semiconductor fabrication. **Darren Williams and Trisha O'Bryon** examine the utility of sessile drop contact angle measurement for surface energy determination and cleanliness verification. They review the available methods, commercial instruments, patents, and literature describing the state-of-the-art in contact angle measurement. This is followed by a description of contact angle measurement techniques that have been modified for use on large surfaces. The negative effects of these changes on accuracy and precision are discussed, and remedies are proposed, including the use of standard reference objects that mimic the size and shape of sessile drops. The combination of these validation tools and the modified contact angle measurement techniques fills a need for robust, production-line-capable cleanliness verification methods.

The contributions in this book provide a valuable source of information on the current status and recent developments in the respective topics on the impact, characterization and removal of surface contaminants. This book will be of value to government, academic, and industry personnel involved in research and development, manufacturing, process and quality control, and procurement specifications in microelectronics, aerospace, optics, xerography, joining (adhesive bonding) and other industries.

We would like to express our heartfelt thanks to all the authors in this book for their contributions, enthusiasm and cooperation. Our sincere appreciation goes to our publishers Frank Hellwig and Matthew Deans, who have strongly supported publication of this volume, in particular, and this series, in general. Renata Corbani and the editorial staff at Elsevier have been instrumental in seeing the book to publication. Rajiv Kohli would also like to thank the staff of the STI library at the Johnson Space Center for their efforts in expeditiously locating obscure and difficult-to-access reference materials.

Rajiv Kohli
Houston, Texas, USA

Kash Mittal
Hopewell Junction, New York, USA

A companion website for this book with all figures including color figures can be found at http://booksite.elsevier.com/9781437778793.

About the Editors

Dr Rajiv Kohli is a leading expert with The Aerospace Corporation in contaminant particle behavior, surface cleaning and contamination control. At the NASA Johnson Space Center in Houston, Texas, he provides technical support for contamination control related to ground-based and manned spaceflight hardware, as well as for unmanned spacecraft. His technical interests are in particle behavior, precision cleaning, solution and surface chemistry, advanced materials and chemical thermodynamics. Dr Kohli was involved in developing solvent-based cleaning applications for use in the nuclear industry and he also developed an innovative microabrasive system for a wide variety of precision cleaning and microprocessing applications in various industries. He is the principal editor of the book series *Developments in Surface Contamination and Cleaning*; the first five volumes in the series were published in 2008, 2010, 2011, 2012, and 2013, respectively, and this book is the sixth volume in the series. Previously, Dr Kohli coauthored the book *Commercial Utilization of Space: An International Comparison of Framework Conditions*, and he has published more than 200 technical papers, articles and reports on precision cleaning, advanced materials, chemical thermodynamics, environmental degradation of materials, and technical and economic assessment of emerging technologies. Dr Kohli was recently recognized for his contributions to NASA's Space Shuttle Return to Flight effort with the Public Service Medal, one of the agency's highest awards.

Dr Kashmiri Lal "Kash" Mittal was associated with IBM from 1972 to 1994. Currently, he is teaching and consulting in the areas of surface contamination and cleaning and in adhesion science and technology. He is the founding editor of the new journal *Reviews of Adhesion and Adhesives* which made its debut in 2013. He cofounded the *Journal of Adhesion Science and Technology* in 1987 and was its editor-in-chief until April 2012. Dr Mittal is the editor of more than 110 published books, many of them dealing with surface contamination and cleaning. He was recognized for his contributions and accomplishments by the worldwide adhesion community which organized in his honor, on his fiftieth birthday, the first International Congress on Adhesion Science and Technology in Amsterdam in 1995. The Kash Mittal Award was instituted in 2002 in his honor for his extensive efforts and significant contributions in the field of colloid and interface chemistry. Among his numerous awards, Dr Mittal was awarded the title of doctor *honoris causa* by the Maria Curie-Sklodowska University in Lublin, Poland in 2003. In 2010, he was honored by both the adhesion and surfactant communities on the occasion of the publication of his 100th edited book.

Contributors

Norman Carnahan, Carnahan Corporation, PO Box 42281, Houston, TX 77242, USA

Rajiv Kohli, The Aerospace Corporation, 2525 Bay Area Boulevard, Suite 600, Houston, TX 77058-1556, USA

Trisha M. O'Bryon, Sam Houston State University, Department of Chemistry, Huntsville, TX 77340-2117, USA

Lirio Quintero, Baker Hughes, 2001 Rankin Road, Houston, TX 77073, USA

Masanobu Sato, Dainippon Screen Mfg. Co., Ltd., Takamiya-cho 480-1, Hikone, Shiga 522-0292, Japan

James T. Snow, DNS Electronics, 2315 Luna Road, Suite 120, Carrollton, TX 75006, USA

Takayoshi Tanaka, Dainippon Screen Mfg. Co., Ltd., Takamiya-cho 480-1, Hikone, Shiga 522-0292, Japan

Darren L. Williams, Sam Houston State University, Department of Chemistry, Huntsville, TX 77340-2117, USA

Chapter 1

Removal of Surface Contaminants Using Ionic Liquids

Rajiv Kohli
The Aerospace Corporation, Houston, TX, USA

Chapter Outline

1. Introduction	2
2. Surface Cleanliness Levels	2
3. Ionic Liquids	4
3.1. Background	5
3.2. Abbreviations and Nomenclature	5
3.2.1. Cations	5
3.2.2. Anions	7
3.2.3. Alkyl Groups (R_n where n = 1, 2, 3, 4, …)	8
3.3. General Characteristics	9
3.4. Thermal Properties	14
3.5. Volatility	16
3.6. Solubility Considerations	17
3.7. Modeling and Predictions of Thermodynamic Properties	18
3.8. Viscosity	19
3.9. Electrical Conductance and High Vacuum Analytical Applications	20
3.9.1. Electron Microscopy	20
3.9.2. Surface Analysis	22
3.10. Toxicity Concerns	23
3.11. Data Compilations	24
3.12. Deep Eutectic Solvents	26
4. Principles of Cleaning with ILs	28
4.1. Basic Principles	28
4.2. Cost Considerations	29
5. Advantages and Disadvantages of ILs	30
5.1. Advantages	30
5.2. Disadvantages	31
6. Applications	32
6.1. Semiconductor Cleaning	32
6.2. Brush Cleaning	33
6.3. Parts Cleaning	33
6.4. Electropolishing	36
6.5. Cleaning with ILs and Supercritical Gases	36
6.6. Cleaning of Oil-Contaminated Sands and Particulate Matter	37
6.7. Decontamination of Hazardous Materials	37
6.8. Microbial Contamination	38
6.9. Cleaning in Place	38
6.10. Cleaning of Artworks	39
6.11. Industrial Applications	39
6.11.1. Hydrocarbon Production	40
6.11.2. Fuel Desulfurization	42
6.12. Consumer Product Applications	42
7. Summary and Conclusions	42
Acknowledgment	43
Disclaimer	43
References	43

1. INTRODUCTION

Conventional solvents, such as chlorinated compounds, hydrochlorofluorocarbons (HCFCs), trichloroethane and other ozone-depleting solvents (ODCs), are commonly used for industrial and precision cleaning in a variety of applications. Many of these solvents are deemed detrimental to the environment [1,2]. Concerns about ozone depletion, global warming, and air pollution have led to new regulations and mandates for the reduction in the use of these solvents. In fact, there is a specific schedule for the United States to phase out its production and consumption of HCFCs in accordance with the terms of the Montreal Protocol as shown in Table 1.1 [3]. The search for alternate cleaning methods to replace these solvents has led to the consideration of various alternative cleaning substances and technologies.

Ionic liquids (ILs)[1] are a new class of materials with unusual and unique properties that make them attractive as process and performance chemicals for a wide range of applications [4–49]. These applications include electrodeposition, electrosynthesis, electrocatalysis, electrochemical capacitor, lubricants, embalming fluids, biocatalysis, plasticizers, solvents, lithium ion batteries, fuel cells, solvents to manufacture nanomaterials, extraction, gas absorption agents, energetic materials for propulsion, and other applications. Figure 1.1 summarizes important properties of ILs and their current and potential applications. Recently, cleaning applications have been proposed and demonstrated. The focus of this chapter is on developments in ILs for removal of surface contaminants.

2. SURFACE CLEANLINESS LEVELS

Surface contamination can be in many forms and may be present in a variety of states on the surface. The most common categories include the following:

- Particles
- Thin film or molecular contamination that can be organic or inorganic
- Ionic contamination
- Microbial contamination.

Other contaminant categories include metals, toxic and hazardous chemicals, radioactive materials, and biological substances that are identified for surfaces employed in specific industries, such as semiconductor, metals processing, chemical production, nuclear industry, pharmaceutical manufacture, and food processing, handling, and delivery.

Common contamination sources can include machining oils and greases, hydraulic and cleaning fluids, adhesives, waxes, human contamination, and particulates. In addition, a whole host of other chemical contaminants from a

---

[1] Throughout this chapter ionic liquids (ILs) and room temperature ionic liquids (RTILs) will be used interchangeably. Most surface cleaning applications employ RTILs.

TABLE 1.1 Comparison of the Montreal Protocol and United States Phase Out Schedules for HCFCs [3]

Montreal Protocol		United States	
Year to be implemented	% Reduction in consumption and production, using the cap as baseline*	Year to be implemented	Implementation of HCFC phase out through Clean Air Act Regulations
2004	35.0	2003	No production and no importing of HCFC-141b
2010	75.0	2010	In addition to the HCFC-141b restrictions, no production and no importing of HCFC-142b and HCFC-22, except for use in equipment manufactured before January 1, 2010 (so no production or importing for NEW equipment that uses these compounds)
2015	90.0	2015	In addition to the HCFC-141b, HCFC-142b and HCFC-22 restrictions, no production and no importing of any other HCFCs, except for use as refrigerants in equipment manufactured before January 1, 2020
2020	99.5	2020	No production and no importing of HCFC-142b and HCFC-22
2030	100.0	2030	No production and no importing of any HCFCs

*The cap for developed countries is set at 2.8% of that country's 1989 chlorofluorocarbon consumption plus 100% of that country's 1989 HCFC consumption.

variety of sources may soil a surface. Typical cleaning specifications are based on the amount of specific or characteristic contaminant remaining on the surface after it has been cleaned.

Most precision technology applications require characterization of particles, as well as nonvolatile residue (NVR). For example, civilian and defense space agencies in the United States (NASA, National Aeronautics and Space Administration; DoD, Department of Defense) and Europe (ESA, European Space Agency)

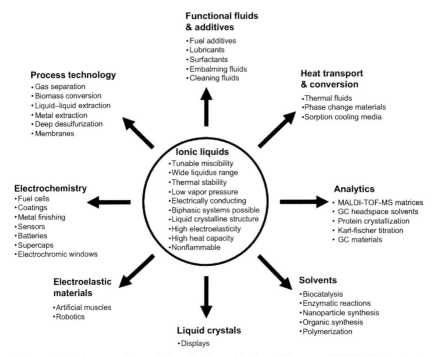

FIGURE 1.1 Current and potential applications of ionic liquids (MALDI-TOF-MS = Matrix-assisted laser desorption/ionization time-of-flight mass spectrometry; GC = Gas chromatography).

specify surface cleanliness levels for space hardware in the microparticle size range [50,51]. The cleanliness levels are based on contamination levels established in the industry standard IEST-STD-CC1246D for particles from Level 1 to Level 1000 and for NVR from Level AA5 (10 ng/0.1 m^2) to Level J (25 mg/0.1 m^2) [52].

The cleanliness levels commonly used by NASA to specify particle and NVR contamination for space hardware are 50 A, 100 A and 300 A (A = 1 mg/0.1 m^2) [50], although for other applications stricter cleanliness levels may be specified, such as Level 10 for particles and Level A/5 (200 μg/0.1 m^2) or A/10 (100 μg/0.1 m^2) for NVR [53]. In many other commercial applications, the precision cleanliness level is defined as an organic contaminant level less than 10 μg/cm^2, although many applications are setting the requirement at 1 μg/cm^2 [50]. These cleanliness levels are either very desirable or required by the function of parts such as metal devices, machined parts, electronic assemblies, optical and laser components, precision mechanical parts, and computer parts.

3. IONIC LIQUIDS

There is a vast amount of published scientific literature on the synthesis, characterization, properties and applications of ILs, represented by Refs [4–49] and the citations therein. No attempt is made to review and summarize the extensive available information except as it relates to surface cleaning. In the following sections, we provide a brief overview of ILs and their characteristics conducive to surface cleaning.

3.1. Background

ILs refer here to purely ionic, salt-like materials that are in liquid form at unusually low temperatures. Broadly defined, ILs are compounds which are liquid below 373 K. More commonly, ILs have melting points below room temperature; some of them even have melting points below 273 K. In general, large and bulky organic cations are combined with weakly coordinating organic or inorganic anions with low-symmetry structures to form the IL. These factors tend to reduce the lattice energy of the crystalline form of the salt and prevent efficient ion lattice packing, resulting in weak coulombic interactions that lower the melting point to give room temperature liquids rather than high-melting solids. Upon mixing, these components turn into liquid at about 313 K or less, and the mixture behaves like an IL.

Some of these salts may have a nitrogen-containing aromatic moiety as the cationic component. Other salts may have a phosphorous-containing cationic component. Typical anionic components of these salts include, but are not limited to, methylsulfate, PF_6^-, BF_4^-, or halides. Table 1.2 lists the cations and anions that have been used to make ILs. Figure 1.2 shows the structure of more common cations and anions.

The cation, anion, and alkyl chain moieties can be adjusted and mixed such that the desired solvating properties, viscosity, melting point, and other properties can be customized for the intended application. These customized ILs are often referred to as "designer solvents".

The combination of a wide range of cations and anions leads to a very large number of possible single-component ILs that could be synthesized, with 10^{12} binary and 10^{18} ternary systems possible. In fact, the number is essentially infinite if we consider quaternary and higher combinations. In reality, the number is of the order of a few thousand ILs that have been described, of which a few hundred are commercially available from suppliers worldwide [54–67].

3.2. Abbreviations and Nomenclature

For the purposes of this chapter, the common cations and anions, and the side chain alkyl groups that constitute ILs are represented as follows [34,68].

3.2.1. Cations

- Imidazolium: IM or im
  - Methylimidazolium: MIM or mim
  - Ethylimidazolium: EIM or eim

TABLE 1.2 Typical Cations and Anions Used for Making Ionic Liquids

Cation	Anion
Imidazoliums	Halide
• Disubstituted imidazoliums, trisubstituted imidazoliums, functionalized imidazoliums, and protonated imidazoliums	• Chloride, bromide, and iodide
Pyridiniums	Borate (e.g. tetrafluoroborate BF_4^-)
• Unsubstituted pyridiniums, substituted pyridiniums, and functionalized pyridiniums	
Ammoniums	Imide
• Symmetrical ammoniums, unsymmetrical ammoniums, functionalized ammoniums, protonated ammoniums, and cholines	
Phosphoniums	Methide
• Symmetrical phosphoniums and unsymmetrical phosphoniums	
Pyrrolidiniums	Sulfate
	• Fluorinated sulfates and non-fluorinated sulfates (e.g. triflate, tosylate)
Piperidiniums	Carbonate
	• Hydrogen carbonate and methylcarbonate
Pyrazoliums	Phosphate (e.g. hexafluorophosphate PF_6^-)
Azoliums	Cyanate
• Triazolium and tetrazolium	• Thiocyanate, dicyanamide, tricyanamethane, and tetracyanoborate
Sulfoniums	Metal halide
Guanidiniums	Carbonic acids
	• Acetate, lactate, salicylate, others (e.g. formate, propionate)
Uroniums and thiouroniums	Others
	• Nitrate, saccharinate, perchlorate, and sulfonate
Specialties	
• Zwitterions and ionic liquid cellulose solutions	

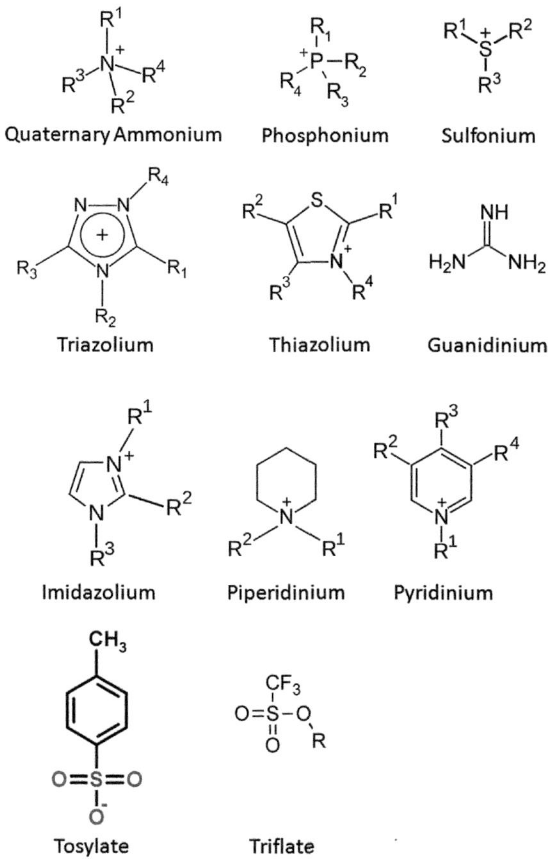

FIGURE 1.2 Structures of typical cations and anions used for making ionic liquids. R is an alkyl group or an aryl group.

- Pyridinium: Py or py
- Pyrrolidinium: Pyrr or pyrr
- Guanidinium: Gu or gu
- Piperidinium: pip
- Phosphonium: P
- Sulfonium: S
- Ammonium: N
- Triazolium: Tz
- Thiazolium: Thia

3.2.2. Anions

- Halides: bromide, Br$^-$; chloride, Cl$^-$
- Nitrate, [NO$_3$]$^-$

- Hexafluorophosphate, $[PF_6]^-$
- Tetrafluoroborate, $[BF_4]^-$
- Alkylsulfate, $[RSO_4]^-$
- Alkylcarboxylate, $[RCO_2]^-$. Acetate $[CH_3CO_2]^-$ is written as [OAc]
- Trifluoromethanesulfonate, $[CF_3SO_3]^-$: triflate is written as [OTf]
- Toluenesulfonate, $[CH_3C_6H_4SO_3]^-$: tosylate is written as [OTs]
- Trifluoromethansulfonylimide, $[N(SO_2CF_3)_2]$: triflimide is written as $[NTf_2]$

3.2.3. Alkyl Groups (R_n where n = 1, 2, 3, 4, ...)

- Ethyl: Et
- Methyl: Me
- Butyl: Bu
- Hexyl: Hx

Alkyl methyl IM is a very common cation and is written as $[C_n\text{mim}]$ or $[C_n\text{MIM}]$ where n is the number of carbon atoms in the linear or branched, substituted or unsubstituted, alkyl, aryl, alkoxyalkyl, alkylenearyl hydroxyalkyl, or haloalkyl groups. As an example, 1-butyl-3-methyl imidazolium cation is denoted as $[C_4\text{mim}]^+$ or $[BMIM]^+$. Similarly, $[C_6py]^+$ denotes 1-hexylpyridinium.

Quaternary ammonium compounds are derivatives of ammonium compounds in which all four of the hydrogens bonded to nitrogen have been replaced with hydrocarbyl groups. Here the symbol C denoting the carbon atom is replaced with N. For example, $[N_{1,8,8,8}]^+$ denotes the methyltrioctylammonium cation. Similarly, for tetraalkylphosphonium cations C is replaced with P as in, for example, $[P_{2,2,2,1}]^+$ which denotes the triethylmethylphosphonium cation. The free electron pairs of one of the two nitrogen atoms in the five-membered imidazoline ring and of the sole nitrogen atom in the five-membered pyrrolidine or six-membered pyridine ring are donated to univalent alkyl groups to produce an N^+ cation.

Trifluoromethanesulfonate, also known by the trivial name triflate, is a functional group. The triflate group is usually represented by [OTf]. For example, 1-triazolium triflate is abbreviated as [Tz1][OTf].

Sulfonium compounds have the structure R_3S^+ cation and an associated anion. As an example, trimethylsulfonium bromide is written $[(CH_3)_3S]$ Br. Generally, but not necessarily, all three R groups are hydrocarbyl.

Polyatomic anions are written with square brackets, but there are no brackets around monoatomic anions. For example, bromide is written as Br^- while dicyanamide is written as $[N(CN)_2]^-$ or abbreviated as [DCA] or [dca].

To represent ILs, the charge signs are deleted when a cation is paired with an anion. Thus, $[C_4\text{mim}][PF_6]$ represents 1-butyl-3-methylimidiazolium hexafluorophosphate.

3.3. General Characteristics

Some of the properties that make ILs attractive alternatives to conventional solvents include the following key features.

1. ILs have a broad liquid range as low as 173 K to as high as 473 K. This feature permits effective process control over a wide range of applications.
2. With a few exceptions, ILs have no measurable vapor pressure; thus, they are easy to handle and they reduce safety concerns where volatility (pollution via an air pathway) could be an issue. This feature also enables vacuum applications.
3. ILs are effective solvents for a broad range of organic, inorganic, and organometallic materials due to their high polarity. This implies low volumes used in cleaning and other process applications.
4. ILs can be tuned to the specific application and chemistry desired. For example, they can be selectively made to have properties ranging from hydrophilic to hydrophobic.
5. ILs are effective Brønsted/Lewis/Franklin acids.
6. ILs are nonflammable below the decomposition temperature.
7. ILs exhibit high thermal stability. Decomposition temperatures above 573 K are not rare.
8. High electrical conductivity of ILs prevents electrostatic charging.
9. High stability of the ILs against oxidation and reduction can be realized.

Typical properties of ILs are compared with the properties of organic solvents in Table 1.3 [23].

Not all ILs are actually liquid at room temperature as shown in Fig. 1.3 [69]. A solid IL, methyl-tri-n-butyl-ammonium dioctyl sulfosuccinate [MeBu$_3$N][DOSS] with a melting point around 313 K is shown on the left, while on the right is 1-butyl-3-methyl imidazolium (diethylene glycol monomethyl ether) sulfate [BMIM][DEGMME SO$_4$] which is a room temperature ionic liquid (RTIL). Figure 1.4 shows an RTIL, [BMIM][NTf$_2$], compared with common table salt. Some ILs are colorless, while others are pale yellow to orange to dark amber, or they exhibit a rainbow of colors for metal-based ILs (Fig. 1.5) [70–84].

ILs can be divided into two broad categories: aprotic ionic liquids (AILs) and protic ionic liquids (PILs). AILs generally consist solely of cations with substituents other than a proton (typically an alkyl group) at the site occupied by the proton in an analogous PIL, and an anion. Examples of AILs are [C$_n$mim][NTf$_2$] and [C$_n$mpyrr][NTf$_2$] families. PILs are produced by proton transfer from a Brønsted acid to a Brønsted base, and are capable of hydrogen bonding, including proton acceptance and proton donation [22,85]. Examples include imidazolium or alkylammonium cations combined with fluorinated or carboxylate anions [86]. There has been increasing interest in PILs for their beneficial characteristics, including low cost, simple synthesis and purification methods, low toxicity and high biodegradability, which tend to outweigh their potentially negative characteristics

TABLE 1.3 Comparison of Typical Characteristics of Representative Organic Solvents with Ionic Liquids

Property	Organic Solvents	Ionic Liquids
Number of solvents	>1000	• 10^{18} theoretically possible • More than 2000 currently described in the published literature • Several hundred commercially available
Applicability	Single function	Multifunction
Catalytic ability	Rare	Common and tunable
Chirality	Rare	Common and tunable
Vapor pressure	Obey the Clausius–Clapeyron equation	Negligible vapor pressure under normal conditions
Flammability	Usually flammable	Usually nonflammable
Polarity	Conventional polarity concepts apply	Polarity concept questionable
Tunability	Limited range of solvents available	Essentially infinite range means "designer solvents" can be created
Viscosity, cP	0.2–100	22–40,000
Density, g/cm³	0.6–1.7	0.8–3.3
Refractive index	1.3–1.6	1.5–2.2
Cost	Normally cheap	Expensive; typically between 2 and 100 times the cost of organic solvents
Environmental impact	Detrimental for many common solvents (high ozone-depletion potential and global warming potential)	Common ILs can be toxic. Tunability enables low toxicity or nontoxic ILs to be designed
Recyclability	Green imperative	Economic imperative

of nonnegligible vapor pressures and slightly lower conductivity than AILs. The first reported room temperature PIL was ethanolammonium nitrate (melting point of 325–328 K) in 1888 [87], followed by hydrazinium azide (melting point of 348 K) and several low melting organic salts in 1891 [88,89], and ethylammonium nitrate (melting point of 285 K) synthesized in 1914 [90].

FIGURE 1.3 Physical appearance of ionic liquids. On the left is methyl-tri-*n*-butylammonium dioctyl sulfosuccinate with a melting point around 313 K. On the right is 1-butyl-3-methyl imidazolium (diethylene glycol monomethyl ether) sulfate which is liquid at room temperature [69]. A color version of this figure appears in the color plate section.

FIGURE 1.4 A room temperature ionic liquid compared with common table salt [47]. A color version of this figure appears in the color plate section.

The key to the unique properties and outstanding flexibility of PILs lies in their chemical nature. PILs are intermediate between fully ILs, such as the widely available dialkylimidazolium salts, and molecular liquids such as hydrocarbons, alcohols and water (Fig. 1.6).

PILs can vary enormously in terms of their properties depending on their position within this "solvent spectrum". PILs formed from the reaction of a strong base with a strong acid are effectively entirely ionic in character and lie toward the right-hand side. These materials characteristically exhibit very low vapor pressures and high ionic conductivities. By contrast, PILs prepared from a weaker base and acid combination exist in a state of constant equilibrium between ionic and molecular species. These tend to exhibit much lower viscosities than fully ILs as well as appreciable vapor pressures and volatility.

The strong ionic (coulombic) interaction within ILs is intentionally weakened to given low melting liquids; nevertheless, these interactions remain

FIGURE 1.5 Metal-based ionic liquids exhibit a wide range of colors. The liquids are from left to right: copper-based compound, cobalt-based compound, manganese-based compound, iron-based compound, nickel-based compound, and vanadium-based compound [84]. *Source: Courtesy of Sandia National Laboratories, Albuquerque, NM.* A color version of this figure appears in the color plate section.

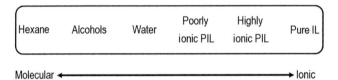

FIGURE 1.6 Protic ionic liquids are located within a "solvent spectrum" from pure ionic liquids to molecular liquids.

strong enough to result in negligible vapor pressure at room temperature in these substances (unless decomposition occurs). The low vapor pressure makes them combustion-resistant, evaporation-proof, and nonflammable, resulting in a highly thermally, mechanically as well as electrochemically stable product, suitable for vacuum applications. In addition, ILs offer other favorable properties: unique and very appealing solvating properties by virtue of their high polarity and charge density, and their immiscibility with water or organic solvents that results in multiphasic systems.

The specific properties of an IL can be almost selected ad hoc, in order to have a compound with the most appropriate characteristics for a specific application. The choice of the cation has a strong impact on the physical properties (melting point, viscosity, conductivity, density, refractive index, etc.) of the IL and will often define its stability, while the chemistry and functionality of the IL is, in general, controlled by the choice of the anion. The anions and cations can be independently selected and combined to design and fine-tune the physicochemical properties of the IL, while at the same time introducing specific features such as controlling solute solubility, hydrophobicity vs hydrophilicity, and other functionalities for a given application. Thus, tailor-made IL materials and solutions are possible.

Although $[C_n mim][PF_6]$ and $[C_n mim][BF_4]$ were the first ILs with specific functionalities—so-called "designer solvents"—and are still dominant

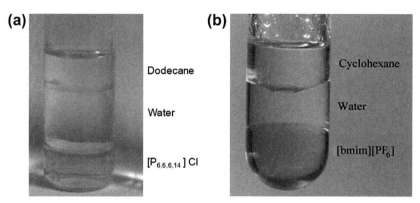

FIGURE 1.7 Examples of triphasic mixtures of ionic liquid, water, and an organic compound [4,95].

in many applications, they have a serious drawback. The $[PF_6]^-$ and the $[BF_4]^-$ anions will degrade in aqueous media, resulting in the formation of toxic and corrosive HF or fluorides. Several non-fluorine-containing ILs have been developed that are stable toward hydrolysis, and are better choices with respect to performance and handling. One example is methyltrioctylammonium thiosalicylate, $[N_{1,8,8,8}][TOS]$ with a melting point below 263 K that has been used to decontaminate various materials [91,92]. This compound is completely stable toward hydrolysis, it is not corrosive, and its constituent ions are nontoxic. Another example is 1,2,3-triazolium salts with various anions (tosylate, triflate, halides, etc.) that exhibit high thermal and chemical stability under alkaline conditions [93,94].

Many ILs are immiscible with water and also do not dissolve alkanes and heavy aromatic compounds. In this case, biphasic or triphasic solutions are formed with the IL (Fig. 1.7(a) and (b)). For example, $[C_n\text{mim}][BF_4]$ salts are miscible with water at room temperature for alkyl chain length less than six, but at or above six carbon atoms, they form a separate phase when mixed with water [4]. This behavior can be of substantial benefit when carrying out solvent extractions or product separations.

This multiphasic behavior has important implications for clean synthesis and for cleaning with supercritical fluids such as supercritical CO_2 (SC-CO_2) [96]. The relative solubilities of the ionic and extraction phases can be adjusted to make the separation as easy as possible. Furthermore, since the IL has effectively no vapor pressure and therefore cannot be lost, volatile products can be separated from the IL by distillation. SC-CO_2 is largely soluble in most ILs, but the solubility of ILs in SC-CO_2 is negligible [97–99]. The combination of the advantages of ILs and SC-CO_2 is a new and interesting application for removal of contaminants. A recent example is the use of [BMIM][PF_4] and SC-CO_2 to clean soils contaminated with naphthalene [100,101].

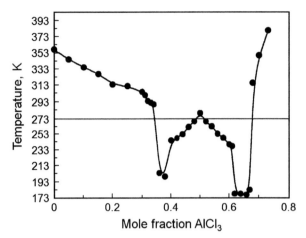

FIGURE 1.8 The phase diagram for mixtures of [EMIM]Cl and AlCl$_3$ [102,103].

3.4. Thermal Properties

A typical IL such as [EMIM][EtSO$_4$] (m.p. 253 K), compared with a typical inorganic salt such as NaCl (m.p. 1074 K), has a significantly lower symmetry, making it more difficult to form a crystal [5]. Furthermore, the charge of the cation as well as the charge of the anion is distributed over a larger volume of the molecule by resonance. As a consequence, the solidification of the IL will take place at lower temperatures. A binary IL system may contain several different ionic species whose melting point and properties depend on the mole fractions of each component. For example, the melting point dependence on composition for the binary IL [EMIM]Cl-AlCl$_3$ is shown in Fig. 1.8 [102,103]. More complex phase behavior has been observed in other IL systems such as [EMIM][Tf$_2$]-AlCl$_3$, alkyl IM, and alkylpyridinium chlorometallate systems [71,73,74,104–111].

By increasing the chain length, the lattice energy of the compound is further reduced and it lowers the melting point (Fig. 1.9). However, there is a maximum chain length before other forms of bonding begin to dominate and a glass transition is observed instead of a melting point (Figs 1.10 and 1.11). In fact, all the temperatures below 273 K are glass transition temperatures rather than true melting points.

The effect of the anion on the melting point is significant. For example, changing from [C$_4$mim]Cl to [C$_4$mim][PF$_6$] or [C$_4$mim][BF$_4$] can change the melting point from 353 K to 278 K or to 202 K, respectively (Figs 1.10 and 1.11), making these lower melting point liquids more fluid and easier to handle.

The lack of a boiling point means that many ILs are liquid over very wide temperature ranges from 300° to 400° from the melting point to the decomposition temperature of the IL.

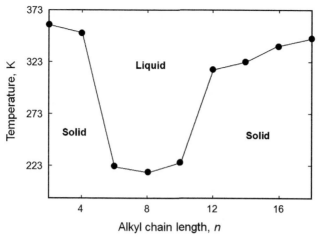

FIGURE 1.9 Melting points for the [$C_n$mim]Cl ionic liquids as a function of the alkyl chain length [5].

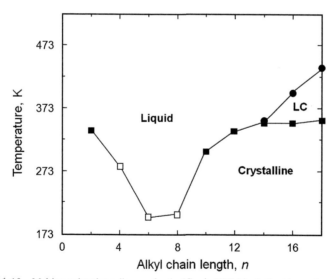

FIGURE 1.10 Melting point phase diagram for the [$C_n$mim][PF_6] ionic liquids as a function of the alkyl chain length. The melting transitions are shown from the crystalline phase (closed squares), glassy materials (open squares) and the clearing transition (closed circles)[2]. LC is the liquid crystal state [5].

---

[2] The liquid crystal state (mesophase) exists within some temperature range, $T_m < T < T_c$, where T_m is the temperature of melting from solid state into a mesophase, and T_c is the clearing temperature, when the liquid crystal transforms into an isotropic liquid. At the clearing temperature T_c, the mesophase melts into an isotropic liquid with no positional and orientational order.

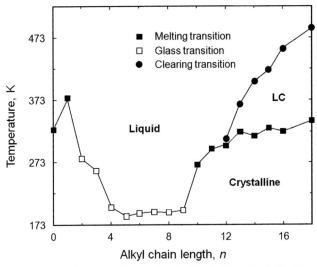

FIGURE 1.11 Melting point phase diagram for the [$C_n$mim][BF_4] ionic liquids as a function of the alkyl chain length. The melting transitions are shown from the crystalline phase (closed squares), glassy materials (open squares) and the clearing transition (closed circles). LC is the liquid crystal state [5].

3.5. Volatility

Volatility of ILs is a key characteristic in the application of ILs for removal of contaminants. Most AILs have nonnegligible vapor pressures and can be distilled under reduced pressures and moderate temperatures without decomposition [112–117], although the vapor pressure under ambient conditions is nonmeasurable [118]. For most processes operating at room temperature, the benefits of nonvolatility of the IL apply. The cohesive energy density of most ILs is very large at room temperature, which is the reason for their low volatility under ambient conditions [119–124]. For higher temperature operations, volatility is a nontrivial consideration and the vapor pressure of the IL is required. Several measurements of the vapor pressure and the enthalpy of vaporization have been reported [38,118,125–140] and various models and methods have been used with varying degrees of success to correlate the data and to predict the vaporization enthalpy [118,129,131,136–149]. Most of the theoretical models are based on atomistic simulations which allow analysis of different contributions to the vaporization enthalpy and also enable parameterization of new force fields [118,144–149]. Recently, quantum chemical methods, such as COnductor-like Screening MOdel for Real Solvents (COSMO-RS), have been successfully employed to obtain reliable predictions of the enthalpy of vaporization of a few imidazolium-based ILs [150], but these calculations are time-consuming and expensive, and usually not straightforward [118]. Even so, the available data are very limited and predictions based on atomistic simulations for new ILs may not be sufficiently validated with experimental data.

3.6. Solubility Considerations

Many organic, inorganic and organometallic materials exhibit high solubility in ILs, which is the basic principle of removal of surface contaminants by ILs. An effective screening tool for selecting the most preferable solvent is the solubility parameter which works by the rule of thumb of "like dissolves like" [151]. Thus, the smaller the difference in solubility parameters between the solute and solvent, the higher the solubility of the solute in the solvent. Thermodynamically, in order for the solvent to be effective in dissolving the solute, the total Gibbs free energy of mixing, ΔG_{mix}, must be zero or negative.

$$\Delta G_{mix} = \Delta H_{mix} - T \Delta S_{mix} \tag{1.1}$$

Here ΔH_{mix}, ΔS_{mix}, and T are the enthalpy of mixing, entropy of mixing, and the absolute temperature, respectively. The solubility parameter, δ, is related to the enthalpy of mixing by the following relationship.

$$\Delta H_{mix} = (v_1 X_1 + v_2 X_2) \times (\delta_1 - \delta_2) \phi_1 \phi_2 \tag{1.2}$$

Here v, X, and ϕ are the molar volume, mole fraction, and volume fraction, respectively; the subscripts 1 and 2 refer, respectively, to the solute and solvent in the mixture.

The solubility parameter is, in turn, defined as the square root of the cohesive energy density, E_D, which is related to the enthalpy of vaporization, ΔH_{vap}, by Eqn (1.3).

$$\delta = (E_D)^{1/2} = [\Delta H_{vap} - RT/v]^{1/2} \tag{1.3}$$

where R is the gas constant.

The δ values for volatile solvents can be obtained directly from ΔH_{vap} or from vapor pressure–temperature data. However, the extremely low vapor pressure of ILs makes it difficult to experimentally measure the ΔH_{vap} values. Thus, both indirect and direct methods have been used for the estimation of solubility parameters from experimental data, including calorimetry, melting temperatures of ILs, inverse gas chromatography, intrinsic viscosity measurements, the activation energy of viscosity, and surface tension measurements [120,121,123,124,152–163].

Various theoretical methods have also been employed to estimate the δ values for ILs, including the Kamlet–Taft equation, nonrandom hydrogen bonding models, statistical associating fluid theory (SAFT), regular solution theory, lattice energy density, molecular dynamics simulations, and group contribution methods [119,144–148,156–166]. Good agreement can be found with the data even though different methods were used to determine the solubility parameters [162,164,165]. However, the available experimental database is still limited to validate estimations of the δ values for new or improved ILs. A shortcoming of most theoretical models is the limited capability to account for strong long-range and directional interactions such as coulombic and hydrogen bonding which can influence the entropy of mixing through various orientational degrees of freedom. However, molecular dynamics simulations can account for most interactions, and have been applied successfully to a few IL systems, including imidazolium, pyridinium, ammonium, phosphonium and guanidinium cations and fluorophosphates, fluoroborate, triflate, and fluoroacetate anions. Still, the best and most accurate predictions of the solubility parameters have been obtained by trial and error [164,167].

3.7. Modeling and Predictions of Thermodynamic Properties

Given the large number of ILs that can be formed, accurate models for the appropriate description of thermodynamic properties of ILs are needed for engineering process applications, in particular the solubility of gases such as CO_2 for cleaning applications. This is important since the solubility of CO_2 in ILs can vary from very low to more than 80% by mole fraction [168,169].

A rigorous molecular thermodynamic model must take into account all the interactions in the system, including hydrogen bonding, electrostatic attraction, ionic and polar interactions, as well as hard chain repulsion and dispersion–attraction forces. For associating and polar molecules, the dimensionless residual Helmholtz energy A^{res} of the system is composed of the sum of the individual contributions of these interactions:

$$A^{res} = A^{hc} + A^{Assoc} + A^{Ion} + A^{Dis} + A^{Polar} \quad (1.4)$$

where the superscripts denote, respectively, the hard chain contribution; the association term due to hydrogen bonding among polar molecules and electrostatic interactions between solvent and solute molecules; coulombic ion–ion interaction; the dispersion term; and the polar term. Each term in Eqn (1.4) is represented by one or more parameters characteristic of each pure component.

Several different theoretical and empirical approaches, correlations and equations of state (EoS) have been used to reproduce experimental data and to predict the thermodynamic properties of new IL systems [144–149,152–162, 165,166,170–178]. These approaches include classical cubic EoS; activity coefficient models; simple and elaborate group contribution methods; square-well chain fluids EoS; lattice models; various SAFT-based models (Soft SAFT, Perturbed Chain (PC)-SAFT, variable potential range SAFT, truncated PC-SAFT (tPC-SAFT), electrolyte (ePC-SAFT), and heterosegmented SAFT). More sophisticated computational techniques include molecular dynamics and Monte Carlo simulations, as well as quantum chemistry calculations (COSMO-RS). The latter approaches are based on well-established theoretical foundations and provide realistic molecular models for pure ILs and their mixtures, and have been applied successfully to several IL systems [144–149,170,172–175]. However, their use requires a lot of experience with specialized software, and the calculations are time-consuming, and expensive. For practical applications of these modeling approaches, tradeoffs must be made among simplicity, accuracy, reliability, time and cost considerations. A simpler approach, such as an EoS, may be warranted, although the lack of a sufficiently large database of measured data tends to limit the usefulness of the EoS approach to reliably predict the thermodynamic properties for new and improved ILs. On the other hand, SAFT-based models have a physical basis for modeling the IL molecules and they fulfill most of the requirements. These models have been the most prevalent among the different modeling approaches to represent gas solubilities in IL systems [98,99,179–189]. Successful examples of their application include the solubility of CO_2, CO, O_2, CHF_3, H_2S, SO_2 and NH_3 in imidazolium-based ILs, H_2 and Xe solubility in [Tf_2N]-imidazolium-based ILs, density and molar volumes of [BF_4]$^-$, [PF_6]$^-$, [NO_3]$^-$, and [Tf_2N]-imidazolium-based ILs, CO_2 solubility and binary vapor–liquid, liquid–liquid, and solid–liquid equilibria in other homologous series of ILs [166,169,172,176–178].

3.8. Viscosity

Generally, ILs are much more viscous than conventional organic solvents, and the viscosity values of most ILs are 2–3 orders of magnitude larger than even heavy organic solvents. For example, the viscosity is only 0.6076 mPa.s for benzene [190], 0.575 mPa.s for toluene [191], 0.42 mPa.s for methyl ethyl ketone [190], or 0.894 mPa.s for cyclohexane [192] at room temperature, whereas it is 70 mPa.s for 1-hexyl-3-methylimidazolium bis(trifluoromethylsulfonyl)imide [193–195] and even as high as 2945 mPa.s and 5647 mPa.s for ethylenediamine di-n-butylphosphate and diethanolamine acetate, respectively [196]. For most cleaning applications with ILs, it is desirable to have low viscosities from a process perspective including reduced power requirements, ease of handling and disposal (dissolution, decantation, filtration, and separation), and high heat and/or mass transfer rates.

For a given cation, the viscosity of the RTILs is strongly determined by the nature of the anion. The viscosity is the lowest for RTILs containing the large [NTf_2]$^-$ anion and the highest for RTILs containing nonplanar highly symmetric or nearly spherical anions. The most viscous ILs are the [PF_6]$^-$-containing salts. In addition, increasing the length of alkyl chains results in higher viscosities because of stronger van der Waals interactions between the larger cations. A significant decrease of viscosity of ILs is generally observed as the temperature increases, represented typically by an Arrhenius-type relationship. The experimental and theoretical information on the viscosity of ILs has been recently reviewed in several publications [20,22,144,147–149,197–204].

The viscosity of ILs is highly sensitive to trace amounts of water and other impurities; even small concentrations of impurities can have a large impact on the viscosity [204–214]. For example, the viscosity of [$P_{6,6,6,14}$][dca] with water content of 300 ppm is around 378 mPa.s compared with around 379 mPa.s for a sample with 249 ppm water [204], indicating that a small increase of 50 ppm water could reduce the viscosity by about 0.4%. The decrease in viscosity is much greater for higher water contents. For example, the viscosity of [C_4mim] [NTf_2] with water content of 19 ppm is 51 mPa.s [211], whereas it is 27 mPa.s with a high water content of 3280 ppm [206], a reduction of nearly 90%. In a few ILs, however, water (5–40% w/w) can increase the viscosity resulting in gel formation in [$C_n$mim]$^+$ cations (n = 8 or 10) with [BF_4]$^-$, halide, nitrate, and other anions [215,216]. This is one reason why it is critical to use high-purity starting materials for making ILs and for their use in cleaning and other process applications. In addition, it is also desirable to develop methods for separation and purification of ILs for recycling and reuse.

Some applications, such as electropolishing, require high ionic conductivities. The high viscosity of the RTILs has a major impact on their conductivities because conductivity is inversely linked to the viscosity. Increasing the length of the alkyl chains generally results in higher viscosity and lower conductivity. The Walden plot of the equivalent molar conductivity (Λ) against the log of the fluidity (inverse viscosity, η^{-1}) is a qualitative way to represent the iconicity of the ILs [14,22,121,199,201,217–224]. Figure 1.12 shows a Walden plot for a series of AILs and for lithium ILs for comparison [221]. The straight line represents the ideal Walden rule, $\Lambda\eta$ = constant, and is calibrated using the data for a 0.01 M KCl aqueous solution, where the ions are known to be fully dissociated and to have equal mobility. However, the average slope of the "ideal" KCl line may not be unity, but is closer 0.87 [222]. Deviations from the reference line in the Walden plot have been used to classify specific ILs as either good or poor ILs, or as nonionic (molecular) liquids [14]. Those ILs having good ionicity are likely to have other related good properties, such as high ionic conductivity. The viscosity and conductivity are often reflected in the glass transition temperature with low values leading to other favorable properties [22]. This can be achieved by modification of the cationic or anionic component of the IL, such as reducing the size of the cation or increasing its symmetry.

3.9. Electrical Conductance and High Vacuum Analytical Applications

ILs can behave as electrically conducting materials. This characteristic along with their extremely low volatility has application in analyses under vacuum conditions. Surface analysis techniques are routinely used to characterize contaminants and the cleanliness of surfaces in precision cleaning applications [225].

3.9.1. Electron Microscopy

Electron beam techniques, such as scanning electron microscopy (SEM) and transmission electron microscopy (TEM), require high vacuum conditions, and often the samples are insulating, or they may be liquid-containing environmental samples. The first direct SEM observation of an IL was a drop of [BMIM][PF_6] (Fig. 1.13) [226]. The dark contrast image indicates that the IL behaves as an electrically conducting material. Furthermore, the IL shows no evidence

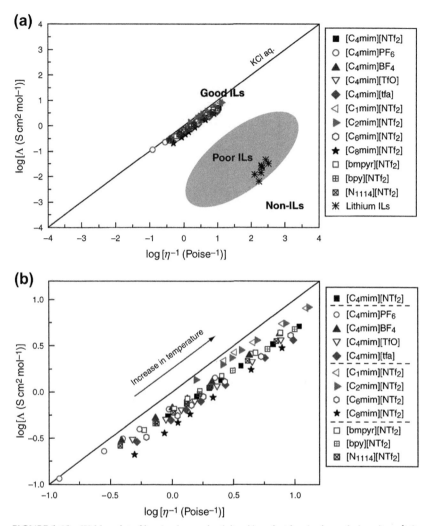

FIGURE 1.12 Walden plot of log (molar conductivity, Λ) against log (reciprocal viscosity η^{-1}) for ionic liquids. The upper figure (a) includes a classification of good and poor ionic liquids, as well as nonionic liquids [14]. The lower figure (b) is a close-up view of the region occupied by typical aprotic ionic liquids. The solid line indicates the ideal line for a completely dissociated strong electrolyte aqueous solution (KCl aq.) [221]. A color version of this figure appears in the color plate section.

of evaporation and maintains its liquid droplet shape under high vacuum conditions and electron beam irradiation. These characteristics are very useful in making insulating materials electrically conducting by coating the specimen with the IL [226,227], or for preparing biological contaminant samples by rehydrating them with ILs [228–232]. Figure 1.14 shows the SEM image of a grain of star sand shell (a type of foraminifer) which was dipped in [EMIM][TFSI]

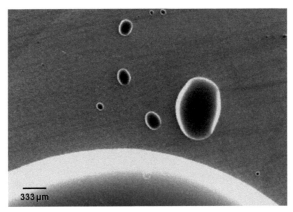

FIGURE 1.13 SEM image of drops of ionic liquid [BMIM][PF$_6$] [226].

FIGURE 1.14 SEM images of two grains of star sand shell. Left grain was dipped in [EMIM][TFSI] and right one was subject to no treatment before observation [226].

(left image), compared with an untreated grain (right image) [226]. The porous shell absorbs and retains the viscous IL. The difference between the images is obvious: the IL-treated grain is a clear image showing details of the surface, whereas the untreated grain gives a highly charged white image due to accumulation of electron charges. Similar experiments have been conducted on different IL-treated materials in a TEM instrument [32,233–237].

3.9.2. Surface Analysis

Surface analysis techniques, such as Auger electron spectroscopy, time-of-flight secondary ion mass spectrometry (TOF-SIMS) and X-ray photoelectron spectroscopy (XPS), can provide detailed information on the chemical structure of surface contaminants [225], but the analysis is performed under high vacuum which is not conducive to wet or liquid samples. The negligible vapor pressure of ILs makes them suitable for investigation by ultrahigh vacuum techniques. A number of studies have been reported using individual and combined surface analysis techniques to investigate ILs [32,238–249]. A major advantage of these techniques, XPS in particular, is that elemental identification is possible and can be used to detect the presence of surface-active contaminants, such as silicone, as well as contaminants that may be dissolved in the IL. Since ILs can be distilled under reduced vacuum (Section 3.5), ultrahigh vacuum technique with line of sight mass spectrometry enables determination of the enthalpy of vaporization of ILs [114,138]. Recently, TOF-SIMS experiments have revealed that a charge pattern could be created on a frozen IL sample surface, and the pattern could be erased by melting the sample [244]. This observation may be applicable to a rewritable data storage system.

Scanning probe techniques are also increasingly being used to investigate various surface and interface phenomena in ILs. Techniques such as atomic force microscopy (AFM) and scanning tunneling microscopy have been successfully applied to probe the structure and image the surfaces in IL/solid systems with atomic level resolution [250–257]. The use of AFM has recently proved to be of major importance in monitoring and characterizing the effectiveness of new enzyme and IL ([BMIM][BF_4] and [EMIM][$EtSO_4$]) formulations for removing protein-based materials from painted and polychrome works of art [257]. This is a demonstration that AFM monitoring protocol can be applied to everyday situations in conservation practice.

3.10. Toxicity Concerns

Toxicity is a major consideration in commercial applications of ILs. If ILs are to be employed on a wide scale in industrial applications in place of volatile organic solvents for removal of surface contaminants, they should possess broader "green" properties including low toxicity and biodegradability. One of the appealing characteristics of ILs for cleaning applications is their nontoxicity in the air environment. However, this is limited mainly to their low vapor pressure under ambient conditions (Section 3.5) which can reduce the risk of air pollution through evaporation or sublimation. On the other hand, most ILs do have finite solubility in water [258–266], and could be released into the aquatic environment via this path through accidental spills, effluents, and other such mechanisms, as well as microbial degradation, sorption and desorption, and similar mechanisms in the terrestrial environment [267–277]. Many of the precursors to ILs are toxic and environmentally hazardous, and many of the more common ILs have toxicities that vary considerably across organisms and trophic levels [37,267–277]. The toxicity does depend on the specific cations and anions. For example, [EMIM]Cl is nontoxic, whereas a close derivative [BMIM]Cl is toxic [278]. This has spurred the development of degradable and biorenewable ILs, as well as the development of many approaches for recycling and recovery of ILs, effluent and wastewater treatments, and other related processes [270,272].

As noted above, many studies have been conducted to assess the toxicity and biodegradation of ILs. However, given the nearly infinite number of possible IL systems and the varied ecosystems, it is infeasible to adequately assess the toxicity of all untested ILs [272]. The current assessment method, quantitative structure–activity relationships (QSARs), is labor-intensive and time-consuming. To overcome this limitation, there has been growing effort to develop QSAR-based models to predict the toxicity of unknown ILs [261,272,279–281]. Recently, a new approach has been devised to screen the toxicity of unknown ILs toward *Vibrio fischeri*, a standard bacterial assay for IL toxicity [282]. This method uses toxicity data of 30 known anions and 64 known cations and the toxicity of toluene or chloroform as a threshold value and applies partial least squares-discriminant analysis (PLS-DA) to screen the 1920 ILs that can be formed by their combinations for ecological toxicity. The accuracy of the model was validated with a test set of 147 samples and achieved a nonerror rate of 93%. The successful results suggest that this model can be used as a screening tool to assist the design of aquatic environmentally friendly ILs and also in development of prediction methods for other organisms and environments.

A simplified approach to visualizing the data from various toxicological tests on ILs has been proposed [283]. In this method, data from each test are given a score ranging from 1 for nontoxic to 3 for very toxic. The scores from the different toxicity tests are then combined to give an overall score for the IL structure. The combined scores are used to visually represent the ILs using a dendrogram (tree diagram) where a score of 1 (nontoxic) is green, 2 (intermediate toxicity) is yellow or amber and 3 (toxic) is red. Each IL is color coded according to its toxicity score and arranged by cation class with increasing alkyl chain length, and by the type of anion. Figure 1.15 shows toxicity dendrograms for selected imidazolium and ammonium-based ILs [276,283]. The dendrogram enables easy visualization of the data to identify toxicity trends and select (nontoxic biocompatible) or rule out (too toxic) ILs for further evaluation.

3.11. Data Compilations

As noted in the previous sections, there is a large body of published experimental data on the thermodynamic and thermophysical properties of ILs. And a variety of empirical and physical correlations and models have been developed to reproduce and predict these properties. Several compilations of the properties of ILs have been published in review articles and in printed books [147,202,284–286], but many of them are for specific properties or for families of ILs. There are a few general databases, such as Reaxys [287] and SciFinder [288], or handbooks, such as Beilstein [289] and Gmelin [290], where a wide range of properties of ILs can be found among a vast array of chemical reaction and substance information in inorganic, organometallic and organic chemistry research. There are also IL databases developed by commercial companies such as Merck, Ionic Liquids Technologies (IoLiTec), and Sigma–Aldrich but these are not complete

(a) Imidazolium salt

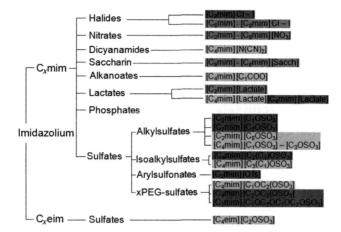

(b) Ammonium salt

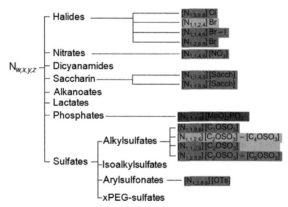

FIGURE 1.15 Proposed dendrograms to represent the toxicities of selected (a) imidazolium and (b) ammonium ionic liquids [276,282]. A color version of this figure appears in the color plate section.

and most of them are for products of the company; some of these databases are not available anymore. Nearly 2000 ILs are included in these compilations for which 29 different physicochemical properties are collected. A large amount of data on the properties of ILs is now contained in web-based searchable comprehensive databases maintained at the National Institute of Science and Technology, Boulder, Colorado, USA (212 ions and 339 ILs as of 2010) [291] and at the Dortmund Data Bank, Dortmund, Germany (766 ILs as of 2011) [292]. Delph-IL is a new web-based IL property database that contains data and synthesis procedures for more than 1000 ILs and it is continuously updated with new property data [293]. One shortcoming of these databases is limited information on IL purity, experimental or simulated determination method, or the source of the ILs (purchased or synthesized). These factors can significantly impact their thermophysical and thermodynamic properties.

Toxicity information on more than 600 ILs is available in the UFT/Merck IL database maintained at the University of Bremen in Bremen, Germany [278].

3.12. Deep Eutectic Solvents

Deep eutectic solvents (DES) are a new generation of solvents that can offset the major drawbacks of common ILs, namely high toxicity, nonbiodegradability, complex synthesis requiring purification, and high cost of the starting materials [294–307]. DES are derived simply by mixing together two safe components (cheap, renewable and biodegradable), which are capable of forming an eutectic mixture. The term DES has been coined mainly to differentiate them from true ILs, and also to reflect the large depression of several hundred degrees in the freezing point of the eutectic mixture (Table 1.4).

TABLE 1.4 Molar Ratio and Temperature at the Eutectic Point in Several Choline Chloride-Based Deep Eutectic Solvents [307]

Deep Eutectic Solvent	Molar Ratio	Eutectic Temperature, K
Choline chloride with urea (Reline)	1:2	285
Choline chloride with ethylene glycol (Ethaline)	1:2	373
Choline chloride with glycerol (Glyceline)	1:2	327
Choline chloride with phenyl-acetic acid	1:2	298
Choline chloride with citric acid	1:1	342
Choline chloride with succinic acid	1:1	344
Choline chloride with malonic acid (Maline)	1:1	283
Choline chloride with oxalic acid (Oxaline)	1:1	307
Choline chloride with m-cresol	1:2	238
Choline chloride with p-cresol	1:2	263
Choline chloride with phenol	1:2	243

The freezing points of the DES are also affected by the organic salt (ammonium or phosphonium salts). For example, urea mixed with ammonium salts in a molar ratio of 2:1 (urea:salt) results in freezing points ranging from 235 K to 386 K for the corresponding DES [307]. The deep freezing point depression of DES is illustrated by the phase diagram of urea–choline chloride system (Fig. 1.16). The melting point of choline chloride is 575 K and of urea is 406 K. But when they are mixed together in a 2:1 (urea:choline chloride) molar ratio, they form an eutectic mixture at 285 K. Typically, the freezing point of a urea–choline salt-derived DES decreases in the order $F^- > [NO_3]^- > Cl^- > [BF_4]^-$, suggesting a correlation with the hydrogen bond strength [307].

For their formation DES need a hydrogen bond donor, such as urea, glycerols, renewable carboxylic acids (for example, malonic, oxalic, citric, succinic or amino acids) or renewable polyols (for example, glycerol, and carbohydrates), in addition to anions and cations. The hydrogen bond donor results in a weaker anion/cation interaction, thereby achieving the low melting temperatures of the DES. One of the most common cations is choline chloride, which is a cheap, biodegradable and nontoxic quaternary ammonium salt. Although DES are made up of cations and anions, they can also be obtained from nonionic species, and are not considered true ILs. Table 1.5 compares the general characteristics of DES with traditional ILs.

The physicochemical properties (density, viscosity, refractive index, conductivity, surface tension, chemical inertness, etc.) of DES, particularly choline chloride-based solvents, are similar to traditional IM-based ILs, and they can be attractive substitutes in many applications. As discussed in Section 6, DES have recently been successfully employed for back-end-of-the-line (BEOL) cleaning in semiconductor applications [308–312] and for other cleaning applications [313,314].

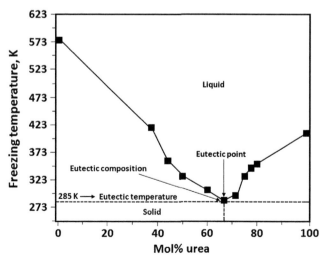

FIGURE 1.16 Phase diagram of the choline chloride–urea system.

TABLE 1.5 Comparison of the Characteristics of Ionic Liquids with Deep Eutectic Solvents

Ionic Liquids	*Deep Eutectic Solvents*
Low-melting point ionic compounds	Low-melting eutectic mixture of compounds
Not always environmental friendly—can be toxic	Biodegradable and nontoxic starting materials
Solution conductivity—moderate to high	Highly conductive
Expensive—recycling is critical	Cheaper than ILs
Highly viscous	Viscosity can be lowered by mixing with suitable ionic solvents
Complex synthesis and purification required	Simple synthesis by mixing inexpensive starting components and no subsequent purification required
Moisture-sensitive ILs must be handled under dry or inert conditions	Generally not moisture-sensitive and convenient storage

4. PRINCIPLES OF CLEANING WITH ILs

4.1. Basic Principles

For cleaning applications, the key attributes of ILs are thermal and chemical stability, melting point, viscosity, solubility, biodegradability, and toxicity, all of which offer advantages compared with other volatile solutions and organic solvents. Their solubility in other solvents and their viscosities can be modified by suitable cation/anion combinations. Solubility control of the ILs is important for the cleaning process, allowing the choice of the specific solvent for removing the contaminant and the IL from the surface. Controlling their viscosity is also important. For example, more viscous ILs will be less penetrative and less disruptive of weathering products during cleaning of sensitive surfaces, such as stained glass in historically significant medieval structures.

One concern regarding the more common ILs is that their balance of intermolecular forces does not match well with the target contaminants in industrial cleaning, making ILs less useful as cleaning solvents [315]. However, this limitation can be overcome by suitable combinations of cations and anions to yield ILs designed for the specific contaminant of interest. Unfortunately, this design feature, in itself, may be a limitation due to limited availability, high cost of synthesis, and lack of toxicological and biodegradability information for new formulations.

4.2. Cost Considerations

Cost is an issue central to any commercial use of ILs [33,62,316,317]. The major cost component in cleaning with ILs is the cost of the IL itself. Pure ILs as available today are manufactured only in kilogram quantities and are not cheap. This high cost stems from (1) the high cost of the components and (2) purification required in the synthesis. Materials to make the dialkylimidazolium cation and fluorine-containing anions are expensive. The price is also influenced by specific requirements on the specification parameters for a particular IL formulation if this makes additional manufacturing steps necessary. Realistically, ILs may never be cheap when compared with other common organic solvents. However, in any cost analysis of a cleaning process the purchase price is only one component. With increasing regulation of volatile organic solvents, ILs with their negligible volatility offer definite cost advantages. As an example, most common solvents such as acetonitrile or benzene can be purchased for less than $1 per kilogram in bulk. Although ILs are unlikely to approach $1 per kilogram, it is reasonable to expect based on economies of scale that price targets of $10–20 per kilogram level could be achieved. This expectation is based on initial production runs carried out at a commercial producer BASF (Ludwigshafen, Germany) on a ton scale.

In general, there is little difference in prices between various ILs. Based on internal access to a very broad range of IL precursors and backward process integration, BASF has been able to perform cost calculations with a high degree of accuracy, indicating only small differences between imidazolium salts and pyridinium, pyrrolidinium, ammonium, and phosphonium salts [61]. The technical performance of the IL will always decide which compound is best suited for the application.

What this means is that ILs would have to be used in applications where they can be recovered and reused. Used solvents are rarely disposed if they have commercial value. For high-priced ILs, recovery and reuse will be essential to economic viability of the process.

Equipment costs for IL cleaning applications are expected to be low, since the ILs are generally used as replacements for existing cleaning solutions or they are additives to the cleaning solutions. New or additional equipment should not be required for an existing process. The chemical inertness of ILs toward most materials means that equipment costs will be comparable to a conventional process even for a new dedicated IL process application. The cost of more conventional solvent process applications also reflects the expenses and capital costs for personnel protection and emission control hardware and monitoring equipment, which may not be required for IL applications. Even if required, they will not be added costs if existing equipment can be used.

The relatively benign nature of ILs suggests that operating costs will be generally low. Most cleaning applications can be performed at ambient temperature and pressure, so energy costs should be low. The major component of the operating cost is IL purification for reuse.

Waste disposal costs for cleaning with ILs are likely to be lower than other solvent cleaning technologies since the waste residue is 100% contaminant. The IL can be recovered and recycled. And if the contaminant can also be recovered, recycled, or reclaimed, there is no cost associated with disposal of waste.

For many cleaning applications, the IL is applied in diluted form. Thus, the incremental cost of the IL itself may not be significant if higher cleaning efficiencies are achieved than competing cleaning processes. On the other hand, undiluted IL used as the cleaning medium will be significantly more expensive in terms of material costs to render the IL cleaning application uncompetitive with other solvent cleaning processes.

The high costs of ILs may be reduced to be competitive with other solvents if DES can be used as an effective replacement for a given cleaning application. As mentioned in Section 3.12, DES can be produced in ton quantities from inexpensive precursors. Also, DES do not require purification, since preparation requires only stirring the components with gentle warming. The purity of the starting materials determines the final purity.

The very limited number of cleaning applications demonstrated with ILs and DES makes it difficult to perform meaningful life cycle cost assessments and comparisons with other solvent cleaning processes at this time. One direct cost comparison has been performed with a water-based cleaning system processing 14,700 kg garments per month [308].

- Water-based system: $0.074/kg per month, which includes costs for water (10%), energy (44%) and cleaning chemicals (46%).
- IL-based system: $0.061/kg per month, which includes costs for the IL solvent (32%), energy (13%) and cleaning chemicals (55%).

The initial high cost of the IL solvent is ameliorated because the solvent is recovered and recycled. Energy costs are lower because no drying is required. Cost of cleaning chemicals is assumed to be the same as in the water-based system, but smaller quantities of chemicals need to be added since the solvent is recycled. In reality the cost of chemicals will be lower. This comparison shows IL-based cleaning processes can offer competitive advantages compared with conventional solvent-based processes.

5. ADVANTAGES AND DISADVANTAGES OF ILs

ILs are being considered and employed for cleaning and related applications. The advantages and disadvantages of ILs for these applications are listed here.

5.1. Advantages

1. ILs offer tunable physicochemical properties for adaptability for a given cleaning application.

2. ILs have high solubility for a very wide range of target contaminants, including organic, inorganic and microbial substances, as well as biomaterials.
3. The tunable properties and high solubility of ILs allow for process intensification and improved cleaning regimes by using low liquid volumes to achieve a given cleanliness level.
4. Chemical inertness toward plant infrastructure makes ILs easy replacements for conventional solvents in existing processes.
5. ILs are largely nonflammable, which is a significant safety advantage in cleaning.
6. Contaminants are the sole waste product. Hence, the waste disposal costs are low. In fact, waste disposal costs may be eliminated if the contaminants can be recovered, reclaimed or recycled.
7. The ILs can be recovered and reused, thus offsetting their high costs.
8. This is a noncorrosive, environmentally friendly process. No hazardous wastes and emissions are generated. ILs and DES can be tuned to be compatible with virtually all materials.
9. Dissolution kinetics is high and cleaning process times are relatively short with ILs, which leads to reduced process operating costs.
10. Energy consumption is expected to be low since there is no heat input to the cleaning process.
11. Cleaning with ILs and DES can be performed at ambient temperatures and pressures which is a significant advantage for cleaning temperature-sensitive parts.

5.2. Disadvantages

1. ILs are expensive, but the prices will be lower as demand increase for large quantities. The high cost makes recovery an economic imperative. The material cost can be offset if cheaper DES of equivalent cleaning effectiveness can be substituted.
2. ILs are highly viscous.
3. Process complexity is high especially if the IL formulation has to be tailored for unknown contaminants. This also requires high level of technical skill.
4. New formulations must be tested to generate information on the chemical and physical properties, biodegradability, and toxicity of the ILs.
5. Many ILs are toxic and require special handling for operation and disposal. Recovery and reuse are also an environmental imperative.
6. There is a very small experience base for cleaning applications with ILs, so overall process operating and life cycle costs are difficult to assess at this time.
7. The cost of purification of used solvents must be considered for recycling and reuse of the IL.

6. APPLICATIONS

ILs and DES have only recently been considered for cleaning applications. Several IL and DES formulations have been proposed and developed for cleaning applications [318–330]. Some examples of IL and DES cleaning applications are discussed below.

6.1. Semiconductor Cleaning

Semiconductor device fabrication involves etching of the photoresist layer to obtain the necessary features on the substrate. Conventional cleaning methods employ strong reactive chemicals to remove the post-etch residues, followed by multiple process steps for rinsing and drying the parts. This is not an environmentally friendly process, requiring large volumes of toxic and corrosive chemicals that must be disposed after use. A new method of cleaning semiconductor substrates has been proposed that overcomes these disadvantages [331]. It uses IL-based cleaning solution compositions for stripping photoresists and cleaning organic and inorganic compounds, including post-etch and post-ash residues, from substrate. High solubility of these residues in ILs allows for process intensification, since the low liquid volumes permit substantial reduction in the amount of chemicals required to achieve the specified cleanliness levels. The cleaning method involves contacting (by immersion or spraying) the contaminated surface of a semiconductor substrate, in the form of a wafer or an integrated circuit, with an IL consisting of an imidazolium, pyridinium, pyrrolidinium, ammonium or phosphonium cation with various halides ($[Cl]^-$), organic ($[OTf]$, and $[OTs]$), and inorganic ($[BF_4]$ and $[PF_6]$) anions. Exposure times are as short as 30 s to 30 min at 293–343 K, depending on the composition of the cleaning solution. The IL can be used undiluted or it can be diluted with other polar solvents. The process results in trace residual contamination below 4 nm and <50 ppb. Because the chemical concentrations and application time can be significantly reduced, more aggressive chemistries can be used for precise process control, resulting in reduced chemical consumption, application of new chemistries for newer semiconductor materials, and significantly reduce or eliminate certain final rinsing and drying steps. The process and chemistries may also have application in nanotechnology device fabrication and in the biotechnology sector.

Recently, DES have been employed for removing post-etch residues formed by CF_4/O_2 etching of resist films on copper in BEOL cleaning in semiconductor processing [309,311,312]. Eutectic mixtures of urea–choline chloride and choline chloride–malonic acid were effective in removing residue films at rates of ~11–17 Å/min and ~30 Å/min, respectively, by immersion cleaning in the temperature range 313–343 K. The higher rate with the choline chloride–malonic acid system can be attributed to the higher solubility of copper oxides in this system [297]. Malonic acid also has high solubility for other metal oxides and is used for decontaminating nickel alloys and stainless steels [313,314].

6.2. Brush Cleaning

Brush cleaning is an established method of removing contaminants (particles, fibers, and other substances) safely and gently from high-value and sensitive surfaces (Fig. 1.17). Unfortunately, electrostatic charging can significantly affect the efficiency of the cleaning process. The process is efficient only if the brush bristles are moistened by a solvent that can also neutralize the charge. Conventionally, dilute NaCl solution has been used that acts as a wetting and antistatic agent to overcome the problem of electrostatic charging. In practice, the brush filaments are moistened by a fine spray of the cleaning solution through a Venturi nozzle. Unfortunately, crystallization of the NaCl from the spray leads to encrustation and blockage of the nozzle and frequent maintenance is required at short intervals. To overcome this problem, quaternary ammonium-based ILs have been used as antistatic additives in the cleaning solutions replacing NaCl [332–335]. The micro-moistened brush filaments effectively bind the contaminants and transport them to the suction system. The effect of replacement on the nozzle is clearly evident (Fig. 1.18) without affecting the performance of the cleaning process.

Another innovative example of the use of ILs (ammonium, oxonium, sulfonium or phosphonium-based salts) as antistatic wetting agents is production of flat structures with antistatic properties [336,337]. Such surfaces can prevent attraction and adherence of dust and other undesirable contaminants. The IL is incorporated in the polymer matrix before curing.

6.3. Parts Cleaning

Several studies have been performed to evaluate the effectiveness of ILs for parts cleaning as replacement cleaners for conventional solvents that are considered ODCs or hazardous air pollutants with high global warming potentials. One study for NASA evaluated 2-ethylhexyl lactate for its efficiency in removing various contaminants (such as hydraulic fluids) that are representative of those encountered in processing oxygen system components. Average cleaning efficiencies of 85% or higher were obtained with stainless steel substrates contaminated with each individual contaminant and their mixtures [338,339]. This cleaning performance was superior to deionized water and was comparable to HFE-7100, which is used for vapor degreasing operations.

Several PIL solutions have been developed for cleaning a variety of contaminated surfaces of metals, ceramics, glasses, semiconductors, and plastic materials [340]. The ILs are obtained by mixing an imidazole diamine with an acid alone, such as acetic or propionic acid, or in combination with acids containing at least one carboxylic function, such as maleic, lactic, succinic or oxalic

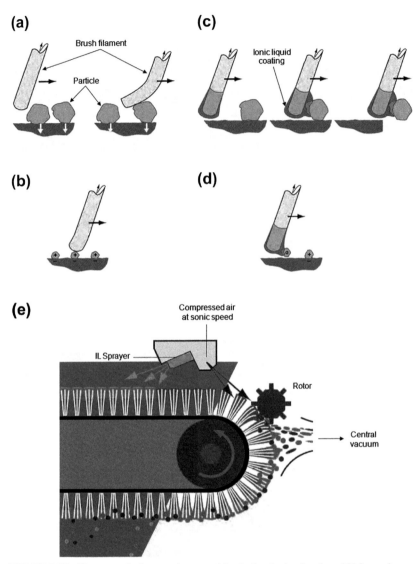

FIGURE 1.17 The removal of contaminant particles by brush cleaning (a and b) is much more efficient if the brush filaments are coated with a conducting ionic liquid film, applied by a spray from a fine nozzle (c, d and e). The adhering particles are removed by a rotor at the end of the process [334,335]. *Source: Courtesy of IoLiTec GmbH and Wandres Micro-Cleaning, Germany.* A color version of this figure appears in the color plate section.

acids. The cleaning solutions are used undiluted, although they can be diluted with water up to 30% by weight. Cleaning is accomplished by immersion in the cleaning tank at an ultrasonic frequency between 30 and 80 kHz in the temperature range 303–343 K. Cleaning efficiencies of 100% were achieved with various combined IL-acid cleaning solution compositions for diverse contaminants

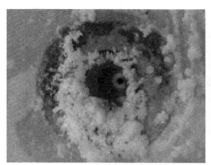 

FIGURE 1.18 A spray nozzle for aqueous solutions of sodium chloride (left) and a hydrophilic ionic liquid (right), each after 10 h of operation [334]. *Source: Courtesy of IoLiTec GmbH, Germany.* A color version of this figure appears in the color plate section.

such as resins, bituminous inks, and cutting oils on silicon wafers, aluminum, copper, stainless steel and molybdenum parts.

In a study for the US Air Force, cleaning efficiency tests using two ILs, [EMIM][Ac] and [EMIM][EtSO$_4$], were conducted on medium carbon, low-alloy steel (Grade 4130 aircraft quality steel) and 2024-T3 aluminum alloy panels coated with MoS$_2$ grease [341,342]. These ILs were selected based on a literature search and vendor survey. In general, [EMIM][Ac] achieved high cleaning efficiencies with both materials, while [EMIM][EtSO$_4$] showed comparable efficiency on the aluminum alloy panels, but lower efficiency on the steel panels. Some chemical etching was observed on the surface of the panels. Chemical properties evaluations and materials compatibility testing showed satisfactory performance. These ILs show potential for use in wipe cleaning applications, although more extensive testing is needed to optimize cleaning performance to ensure they meet Air Force requirements.

Hot processing of metals into finished shapes, such as strip and wire, results in the formation of oxide scale on the surface. These oxides must be removed from the finished metal product which has to be bright metallic and clean with no oxide residues on the metal surface. Conventional scale removal methods involve chemical, mechanical, or thermal treatments which have serious disadvantages, including use of corrosive acids and hazardous chemicals, dust and fine particulate generation, loss of metal, and high temperatures. A new IL-based method has been developed and successfully demonstrated for single-step removal of oxide scale on parts (sheets, coils, strips, wires, and rods) made from various grades of steels [324]. This method employs an IL composition that can perform conditioning and pickling of the oxide scale on metal parts, followed by removal of the scale. The IL is composed of (1) at least one organic salt as a source of inorganic Lewis-basic anions (such as halides or pseudohalides of organic cations) and (2) one or more Lewis-acidic inorganic metal salts of which at least one metal salt comprises a metal cation with an oxidation state more positive than the lowest positive oxidation state of the metallic element itself (such as FeCl$_3$). The molar ratio of components a:b is less than 1:1. Removal of the scale is accomplished by contacting the part with the IL composition for a short time (typically 30 min) at 373 K for open-treatment baths or at lower temperatures under vacuum.

An intriguing observation that may have application to removal of surface contaminants is that an IL droplet rolling off on an inclined solid surface collects all surface impurities along its path, leaving no microscopically visible traces behind [343,344]. Since ILs are powerful solvents, it is expected that such cleaning action will also remove grease spots from the surface.

6.4. Electropolishing

A large variety of metallic substrates, such as stainless steels, molybdenum and titanium alloys, aluminum, and nickel–cobalt alloys, can be electropolished in ILs and DES to yield a clean, shiny, and smooth surface (Fig. 1.19) [320,345–347]. The high solubility of oxides in ILs and DES allows for process intensification [297,348]. The operating parameters are comparable to those in existing acid-based electrolytes, but significantly higher current efficiency can be achieved. Typical operating parameters include a process temperature of 303–323 K and 3–5 V for 10 min. The polishing process can be improved significantly by additives such as oxalic acid, enabling the process to be extended to other systems. The major advantages of the process include use of noncorrosive electrolyte solution, improved surface finish, reduced and simplified waste management, recycling and reuse of the electrolyte, and safer operating conditions.

6.5. Cleaning with ILs and Supercritical Gases

A novel application of IL and SC-CO$_2$ is to serially clean contaminated soils [100,101]. The IL is used to dissolve soil contaminants under ambient conditions and SC-CO$_2$ is used to recover the contaminants from the IL extracts. The efficacy of the process has been demonstrated by extracting naphthalene from soil with [BMIM][PF$_6$] IL. The amount of naphthalene remaining in the

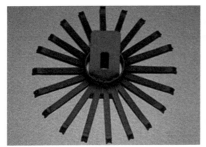

FIGURE 1.19 Surface finish of a titanium alloy part electropolished in choline chloride-based ionic liquid (right image) compared with the original surface (left image) [345,347].

soil (~21–44 µg) was below the contamination limit of 50 µg. Subsequently, SC-CO_2 was used to recover the naphthalene dissolved in the IL. At 313 K and 14 MPa, naphthalene recovery was nearly 84% for an extraction time of 4 h. A process flow sheet has been developed for IL extraction of contaminated soils and continuous SC-CO_2 extraction of the contaminants from IL extracts for recovery and reuse of the IL.

Recently, ILs have been proposed as detergents to improve the effectiveness of the dry-cleaning process using SC-CO_2 [349]. The IL additive helps remove hydrophilic and/or polar impurities more effectively than conventional detergents that are ineffective in removing such impurities.

6.6. Cleaning of Oil-Contaminated Sands and Particulate Matter

A new IL-based method of separating hydrocarbons from tar and beach sands and particulate matter has been developed [350,351]. This method uses IM-based ILs, or other ILs that are soluble in water and insoluble in nonpolar organic solvents, to separate the heavy viscous oil from the sand, as well as separating oil from drill cuttings. By simply mixing the components together at room temperature, three-phase separation is achieved into minerals, IL, and hydrocarbons layers without generating wastewater. Standard solid–liquid and liquid–liquid extraction techniques can be used to separate the components. Bitumen is easily separated from the tar sands, but separation of the residual oil from the drill cuttings requires the addition of a volatile solvent to the mixture. The solvent can be subsequently separated from the oil by vacuum distillation. This method uses very little energy and water, and all solvents are recycled and reused. The method has also been demonstrated on the beach sand contaminated by the Deepwater Horizon oil spill in the Gulf of Mexico. Complete separation of the tar from the sand could be achieved with a separation solution containing [EMIM]Cl.

6.7. Decontamination of Hazardous Materials

Hazardous materials in the form of wastes are a toxic by-product of industrial activity in the public and private sectors. Conventional methods of destroying chlorinated hydrocarbon waste include thermal incineration, supercritical water oxidation, direct chemical oxidation, photochemical oxidation, and solvent extraction among many different treatment techniques [352]. Each of these techniques has disadvantages and limitations, including high capital, installation, energy, and operating costs, public acceptability, handling and disposal of the process wastes, and not environmentally benign. A new process has been proposed for the destruction of halogenated hydrocarbons using a superoxide ion in DES [353,354]. The superoxide, such as H_2O_2, can be generated electrochemically by reduction of oxygen in the DES or dissolving alkali or alkaline-earth superoxides in the DES. A wide variety of halogenated hydrocarbons and chemical warfare agents were destroyed by reacting them with the DES at ambient pressure and temperature. No toxic by-products were produced.

An alternative treatment for decontamination and detoxification of chemical warfare agents is based on surfactant microemulsions that incorporate various ILs with tailored properties [355,356].

6.8. Microbial Contamination

Microbial contamination is a subject of growing concern in the health and food sectors. The risk of transmission of diseases, chronic plant, animal and human infections, failure of medical implants and allograft tissue, and microbial corrosion and biofouling are some of the disastrous consequences caused by microbial contamination. Many ILs display strong activity against clinically significant Gram-positive and Gram-negative bacteria, algae, and fungi [268–277,357–363]. They are also effective in breaking down microbial biofilms that cause infections in hospitals and other medical facilities [361]. By altering the cation and anion pairings and the alkyl chain lengths, the toxicity and other properties of the ILs can be tuned, thereby facilitating their use as surface biocides for disinfecting contaminated surfaces. For example, methyl and hydroxyethyl-substituted IM salts show significantly improved antimicrobial potency of these ILs [269,357]. ILs are being proposed as biocides, disinfectants, antiseptics, or sterilization agents for medical devices and instruments, or as preservatives. The ILs could be applied to a surface that is already contaminated, or the surface could be coated to prevent biofilms from forming, such as in anti-infective medical devices. ILs may also have application as antifouling agents in various industrial and marine applications where biofilms can clog pipes, cause surface corrosion, and affect the performance of the systems.

6.9. Cleaning in Place

The contamination of equipment by both products and by-products is a major problem in pharmaceutical manufacturing. The ability of ILs to dissolve large concentrations of pharmaceutical compounds and intermediates that are otherwise intractable to removal by conventional solvents is a benefit to the pharmaceutical process industry. For example, ammonium-based ILs have been shown to dissolve 300–500 g/L of amoxicillin [364], a contaminant in pharmaceutical manufacturing. The solubility of amoxicillin in conventional organic solvents is low, making the cleaning process less effective and requiring handling and disposal of large volumes of the used solvents. Cleaning with ILs offers the ability to selectively dissolve specific compounds and residues, high solubility of the target contaminants, noncorrosive solvent exposure of the equipment, reduced waste disposal costs, and user safety and convenience. Cleaning can be effected in place by spraying or immersing the contaminated equipment in the cleaning solution.

6.10. Cleaning of Artworks

Medieval stained glass has relatively high content of alkali and alkaline-earth ions (potassium, sodium and calcium) and low percentage of silica. This composition contributes to its deterioration due to relative humidity fluctuations, pollutants, and biological activity. In the presence of atmospheric gases such as CO_2 and SO_2, corrosion crusts are formed on the surface, composed mainly of calcium salts ($CaCO_3$ and $CaSO_4$, as well as CaC_2O_4) that form from the interaction between the glass and oxalic acid produced by microorganisms at high humidity levels (>85%) [365]. Conventional mechanical and chemical cleaning methods used by conservators can be aggressive to the glass, or they have low efficiency in the removal of the corrosion crusts. They can also damage the weathering products that are considered part of the historical record of the glass.

In a recent study [366,367], the use of ILs was assessed as alternative to conventional mechanical and chemical cleaning methods on the fifteenth century stained-glass panel S07c, Figura Aureolada, from the Monastery of Santa Maria da Vitória, in Batalha, Portugal. The ILs were selected from quaternary imidazolium, phosphonium, and quaternary ammonium cations, combined with chloride, dicyanamide and ethyl sulfate anions. [EMIM][EtSO4] and [OMIM][DCA] revealed the best results and were used for cleaning the two stained-glass fragments from the panel S07c. Their main action was to soften the corrosion layer. In the case of the two fragments in question the results were successful, proving that it was possible to remove the crusts in a controlled way. The experiments suggest that ILs are a good alternative to current cleaning methods applied to stained glass, but further testing is needed before ILs are considered safe for cleaning stained glass.

In another recent example, new formulations of enzymes (proteases) and ILs [BMIM][BF4] and [EMIM][EtSO4] have been used to successfully remove proteinaceous varnish layers (egg white, animal glue, isinglass and casein) from painted and polychrome works of art [368]. The treatment was applied on documented reconstruction of paintings and gold leaf gilding. These innovative IL + enzyme formulations suggest the use of ILs as an alternative solvent to enzymes alone (proteases, amylases, lipases, and cellulases) [369,370]. A key factor in enzyme cleaning of paintings is the outdoor temperature which can significantly slow the cleaning process. The cost of the enzymes is also a reason for the lack of their widespread use. ILs can be designed (cation–anion combinations) to meet different requirements such as improvement of enzyme cleaning rate and effectiveness, surface compatibility, and safety.

6.11. Industrial Applications

Some industrial contamination-related applications are discussed below.

6.11.1. Hydrocarbon Production

Scale formation is a common reason for a decline in hydrocarbon production in or on the wellbore and in the near-wellbore region of the hydrocarbon-bearing formation. The scale is composed of minerals, such as $BaSO_4$, $CaCO_3$, and $CaSO_4$, which may build up to eventually choke off the wellbore. Common methods for scale removal are mechanical or chemical techniques, which are often ineffective or damaging to the steel lining of the wellbore. A new method [371] for scale removal in wellbores exploits one or more attributes of ILs.

1. Dissolution of the scale due to the high solubility of a wide range of organic and inorganic materials in the selected IL. Figure 1.20 compares the dissolution rate of $BaSO_4$ in trimethylamine dialuminum heptachloride (TMA-HIL-67) with a conventional ethylenediaminetetraacetic acid-based solvent.
2. Substantial heat is generated on IL formation from the precursors pumped downwell separately and allowed to react in the wellbore or near-wellbore area. The heat generated can melt the contaminants (paraffin, wax, and sludge) coating the wellbore. The IL is then directly in contact with the scale and can dissolve it. The precursors can be encapsulated in a permeable membrane for delayed or controlled release.
3. ILs can act as a carrier solvent to transport a variety of agents or materials, including scale-removal agents, such as strong or super acids to the scale deposit(s).
4. Reaction of the selected IL with an aqueous liquid to generate acids or super acids that can react with the scale. The rate of acid generation can be controlled by controlling the rate of water addition. The efficiency of acid generation by the selected IL composed of an $[AlCl_4^-]$ anion is much higher than organic acids, allowing for process intensification.

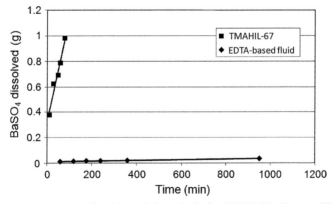

FIGURE 1.20 Dissolution of $BaSO_4$ scale in an ionic liquid TMAHIL-67 vs an EDTA-based solvent [371].

The results from laboratory testing demonstrate that ILs are suitable candidates for the dissolution of multiple types of scale common in oil field environments, particularly $BaSO_4$, $CaSO_4$, and $CaCO_3$ [371].

Drilling fluids, commonly used in drill pipes for hydrocarbon recovery or for geothermal energy, serve multiple functions, including cooling the rotary drilling bit, suspend and remove the borehole cuttings, act as barrier to prevent entrance of the fluids into the surrounding earth, and counteract the pressure from the surrounding earth formation. Additives such as $CaCO_3$, Fe_2O_3 or $BaSO_4$ are incorporated in the fluid to enhance many of these functions. Unfortunately, these additives could separate precipitate from the fluid, exacerbating scale formation and leading to interruption or stoppage of drilling operations. Various compositions of ILs have been proposed as replacements for the conventional oil-based drilling fluids to mitigate these shortcomings [372]. The IL-based fluids are composed of a single IL or a mixture of ILs, and are designed to meet specific fluid property requirements for the given drilling application. Furthermore, the IL fluid can be recovered, cleaned, and reused.

Fracturing is a new technology for enhanced recovery of hydrocarbons. In this process, cracks are formed by drilling in the subterranean strata. Cellulosic materials are frequently added to thicken the drilling or fracturing fluid for enhanced drilling performance, but in the presence of water, these materials hydrolyze to revert back to cellulose. However, cellulose is sparingly soluble in water and most organic solvents, requiring frequent cleaning of the wellbore and resulting fractures to remove deposited cellulosic material that can impede or stop the flow of hydrocarbons through the fractures. On the other hand, RTILs such as [C_4mim]Cl can dissolve up to 25 wt% cellulose, which is considerably higher than organic solvents [373–376]. As noted previously, however, ILs are expensive and many salts are toxic, and require special handling. By comparison, DES are produced from inexpensive, nontoxic precursors, and can be handled safely (Section 3.12). A new method using DES has been developed for solubilizing/removing cellulose or chemically modified cellulosic materials utilized in subterranean drilling operations [328]. DES useful as cellulose solvents include quaternary ammonium compounds, including choline chloride and chlorocholine chloride, reacted with a compound selected from amides, amines, carboxylic acids, alcohols, and metal halides. The method involves pumping the DES downhole after fracturing operations to remove cellulosic material left behind in the fractures, on the face of the formation, along the wellbore, or elsewhere within the subterranean region. The DES can be used alone, or in sequential treatment, whereby DES treatment is followed by flushing with water, caustic or acid solutions, or anhydride substances. The tested DES were capable of solubilizing up to 50 wt% cellulosic material such as xanthan gum, cellulose fibers, modified guar gum, carboxymethyl tamarind, and sodium carboxymethyl cellulose. By comparison, the [C_4mim]Cl IL dissolved a maximum of 25 wt% cellulosic material.

6.11.2. Fuel Desulfurization

Another cleaning-related application is desulfurization of fuels using various classes of ILs. Several variations of the process have been reported and have been reviewed recently [377,378]. This involves direct liquid–liquid extraction of sulfur-containing compounds in a single step, or oxidative desulfurization in which the sulfur compounds are extracted and oxidized to sulfoxides and sulfones in the extractant, followed by liquid–liquid extraction to more efficiently remove the compounds. IL desulfurization technologies have several limitations that must be overcome before commercial implementation is feasible. Most of the reported studies involved model fuels under laboratory conditions that are not representative of real-world conditions. Although the reported extraction efficiencies were high, preparation of efficient catalysts is still a complicated process and not commercially available. Other limitations relate to the partitioning of the components, need for multiple extraction cycles, loss of hydrocarbons, and decomposition of the ILs on exposure to water.

6.12. Consumer Product Applications

Many ILs are biodegradable and nontoxic liquids, which can be used in consumer product applications. They also exhibit the ability to dissolve a wide variety of materials and contaminants including salts (lime scale, bleaches, and metal tarnishes); fats; proteins and amino acids; anionic, cationic and nonionic surfactants that can then be used to solubilize oils; sugars and polysaccharides; and metal oxides and a wide range of solutes. For many of these ILs, the production costs are low associated with the inexpensive raw materials used for their production. These attributes make them suitable for use as cleaners (liquid or aerosol sprays, pour-on liquids, and liquid additives), both in the household and in an industrial environment, as well as for other consumer applications [317,319,322,325–327,349,379–381]. The IL-containing compositions can be formulated in the form of liquid, gel, paste, foam, or solid. Some examples of consumer product applications are detergent formulations for laundry, dish washing, and hard surface cleaning; biofilm removal; waterless garment cleaning; textile, yarn, and leather cleaning with liquid or supercritical CO_2; fabric care products; interior and exterior car care products; personal care products such as hair coloring and conditioning and baby shampoo formulations; and jewelry cleaning.

7. SUMMARY AND CONCLUSIONS

ILs and DES are a new class of low melting point solvents with attractive characteristics for cleaning applications. These characteristics include high solubility for a wide range of contaminants, thermal and chemical stability, low melting point (even below 273 K), low volatility, and high conductivity. IL solvents do have some disadvantages including high cost, complex synthesis, need for purification for reuse, and high toxicity and nonbiodegradability of several formulations. Several DES formulations have been developed to overcome some of these limitations. Cleaning applications have been successfully demonstrated although many of these are largely still at the laboratory stage. Applications range from semiconductor wafer and integrated circuit cleaning, brush cleaning for microcontaminants, precision cleaning of parts for aerospace applications, oxide-scale removal on metals, electropolishing of metals, microbial decontamination, cleaning of artworks, cleaning of wellbores in oil and gas recovery, soil decontamination, and consumer product applications.

ACKNOWLEDGMENT

The author would like to thank the staff of the STI Library at the Johnson Space Center for help with locating obscure reference articles.

DISCLAIMER

Mention of commercial products in this chapter is for information only and does not imply recommendation or endorsement by The Aerospace Corporation. All trademarks, service marks, and trade names are the property of their respective owners.

REFERENCES

[1] U.S. EPA, The U.S. Solvent Cleaning Industry and the Transition to Non Ozone Depleting Substances, EPA Report U.S. Environmental Protection Agency (EPA), Washington, D.C., 2004. www.epa.gov/ozone/snap/solvents/EPASolventMarketReport.pdf.
[2] J.B. Durkee, Cleaning with Solvents, in: R. Kohli, K.L. Mittal (Eds.), Developments in Surface Contamination and Cleaning, William Andrew Publishing, Norwich, NY, 2008, pp. 759–871. (Chapter 15).
[3] U.S. EPA, HCFC Phaseout Schedule, U.S. Environmental Protection Agency, Washington, D.C., 2012. http://www.epa.gov/ozone/title6/phaseout/hcfc.html.
[4] M.J. Earle, K.R. Seddon, Ionic liquids. Green solvents for the future, Pure Appl. Chem. 72 (2000) 1391.
[5] D.W. Rooney, K.R. Seddon, Ionic Liquids, in: G. Wypych (Ed.), Handbook of Solvents, ChemTec Publishing, Toronto, Canada, 2001, pp. 1459–1484. (Chapter 21.2).
[6] R.J. Lempert, P. Norling, C. Pernin, S. Resetar, S. Mahnovski, Next Generation Environmental Technologies. Benefits and Barriers, Appendix A17. Room Temperature Ionic Liquids, Rand Monograph Report MR-1682-OSTP Rand Corporation, Santa Monica, CA, 2003. www.rand.org.
[7] Z. Mu, W. Liu, S. Zhang, F. Zhou, Functional room-temperature ionic liquids as lubricants for an aluminum-on-steel system, Chem. Lett. 33 (2004) 524.
[8] P. Majewski, A. Pernak, M. Grzymisławski, K. Iwanik, J. Pernak, Ionic liquids in embalming and tissue preservation: can traditional formalin-fixation be replaced safely? Acta Histochem. 105 (2005) 135.

[9] A. Pernak, K. Iwanik, P. Majewski, M. Grzymisławski, J. Pernak, Ionic liquids as an alternative to formalin in histopathological diagnosis, Acta Histochem. 107 (2005) 149.
[10] J. Pernak, F. Stefaniak, J. Węglewski, Phosphonium acesulfamate based ionic liquids, Eur. J. Org. Chem. 4 (2005) 650.
[11] C.L. Liotta, P. Pollet, M.A. Blecher, J.B. Aronson, S. Samanata, K.N. Griffith, Ionic liquid energetic materials, U.S. Patent Application 2005/0269001, 2005.
[12] Sigma-Aldrich, Enabling Technologies – Ionic Liquids, Company publication, Sigma-Aldrich, St. Louis, MO, 2005, ChemFiles, vol. 5, No. 6, pp. 1-23 . http://www.sigmaaldrich.com/content/dam/sigma-aldrich/docs/Aldrich/Brochure/al_chemfile_v5_n6.pdf.
[13] Sigma-Aldrich, Ionic Liquids, Company publication, Sigma-Aldrich, St. Louis, MO, 2006, ChemFiles, vol. 6, No. 9, pp. 1-18 . http://www.sigmaaldrich.com/content/dam/sigma-aldrich/docs/Aldrich/Brochure/al_chemfile_v6_n9.pdf.
[14] C.A. Angell, N. Byrne, J.-P. Belieres, Parallel developments in aprotic and protic ionic liquids: physical chemistry and applications, Acc. Chem. Res. 40 (2007) 1228.
[15] J.F. Brennecke, R.D. Rogers, K.R. Seddon (Eds.), Ionic Liquids IV. Not Just Solvents Anymore, ACS Symposium Series, vol. 975, American Chemical Society, Washington, D.C., 2007.
[16] P. Wasserscheid, T. Welton, Ionic Liquids in Synthesis, second ed., Wiley-VCH, Weinheim, Germany, 2007.
[17] A.P. Abbott, D.L. Davies, G. Capper, R.K. Rasheed, V. Tambyrajah, Ionic liquids and their use, U.S. Patent 7,196,221, 2007.
[18] J.M. Shreeve, Ionic Liquids as Energetic Materials, Report AFRL-SR-AR-TR-07-0094, Air Force Research Laboratory, Air Force Office of Scientific Research, Arlington, VA, 2007.
[19] M. Stasiewicz, E. Mulkiewicz, R. Tomczak-Wandzel, J. Kumirska, E.M. Siedlecka, M. Gołebiowski, J. Gajdus, M. Czerwicka, P. Stepnowski, Assessing toxicity and biodegradation of novel, environmentally benign ionic liquids (1-alkoxymethyl-3-hydroxypyridinium chloride, saccharinate and acesulfamates) on cellular and molecular level, Ecotoxicol. Environ. Saf. 71 (2008) 157.
[20] A.S. Pensado, M.J.P. Comuñas, J. Fernańdez, The pressure–viscosity coefficient of several ionic liquids, Tribol. Lett. 31 (2008) 107.
[21] M. Koel (Ed.), Ionic Liquids in Chemical Analysis, CRC Press, Taylor & Francis Group, Boca Raton, FL, 2008.
[22] T.L. Greaves, C.J. Drummond, Protic ionic liquids: properties and applications, Chem. Rev. 108 (2008) 206.
[23] N.V. Plechkova, K.R. Seddon, Applications of ionic liquids in the chemical industry, Chem. Soc. Rev. 37 (2008) 123.
[24] J.F. Wishart, Energy applications of ionic liquids, Energy Environ. Sci. 2 (2009) 956.
[25] I. Minami, Ionic liquids in tribology, Molecules 14 (2009) 2286.
[26] M.D. Bermúdez, A.E. Jiménez, J. Sanes, F.J. Carrión, Ionic liquids as advanced lubricant fluids, Molecules 14 (2009) 2888.
[27] N.V. Plechkova, R.D. Rogers, K.R. Seddon (Eds.), Ionic Liquids: From Knowledge to Application, ACS Symposium Series, vol. 1030, American Chemical Society, Washington, D.C., 2009.
[28] B. Kirchner (Ed.), Ionic Liquids, Topics in Current Chemistry 290, Springer Verlag, Berlin and Heidelberg, Germany, 2009.
[29] F.M. Kerton, Room Temperature Ionic Liquids and Eutectic Mixtures, Alternative Solvents for Green Chemistry, RSC Publishing, Oxford, UK, 2009, pp. 118–142. (Chapter 6).
[30] R. Giernoth, Task-specific ionic liquids, Angew. Chem. Int. Ed. 49 (2010) 2834.

[31] C. Chiappe, M. Malvaldi, Highly concentrated "solutions" of metal cations in ionic liquids: current status and future challenges, Phys. Chem. Chem. Phys. 12 (2010) 11191.
[32] T. Torimoto, T. Tsuda, K.-I. Okazaki, S. Kuwabata, New frontiers in materials science opened by ionic liquids, Adv. Mater. 22 (2010) 1196.
[33] J. Gorke, F. Srienc, R. Kazlauskas, Toward advanced ionic liquids. Polar, enzyme-friendly solvents for biocatalysis, Biotechnol. Bioprocess Eng. 15 (2010) 40.
[34] M. Freemantle, An Introduction to Ionic Liquids, Royal Society of Chemistry, Cambridge, UK, 2010.
[35] Y. Zhang, H. Gao, Y.-H. Joo, J.M. Shreeve, Ionic liquids as hypergolic fuels, Angew. Chem. Int. Ed. 50 (2011) 9554.
[36] T. Hawkins, R&D of Energetic Ionic Liquids, Presentation at Partners in Environmental Technology, Washington, D.C., December 2011. http://www.dtic.mil/cgi-bin/GetTRDoc?AD=ADA554403.
[37] H. Gao, J.M. Shreeve, Azole-based energetic salts, Chem. Rev. 111 (2011) 7377.
[38] K.W. Street Jr., W. Morales, V.R. Koch, D.J. Valco, R.M. Richard, N. Hanks, Evaluation of vapor pressure and ultra-high vacuum tribological properties of ionic liquids, Tribol. Trans. 54 (2011) 911.
[39] A. Kokorin (Ed.), Ionic Liquids: Theory, Properties, New Approaches, InTech Publishing –Open Access Company, Rijeka, Croatia, 2011. www.intechopen.com/books/ionic-liquids-theory-properties-new-approaches.
[40] S.T. Handy (Ed.), Ionic Liquids – Classes and Properties, InTech Publishing – Open Access Company, Rijeka, Croatia, 2011. http://www.intechopen.com/books/ionic-liquids-classes-and-properties.
[41] A.A.J. Torriero, M.J.A. Shiddiky (Eds.), Electrochemical Properties and Applications of Ionic Liquids, Nova Science Publishers, Hauppauge, NY, 2011.
[42] Ch. Reichardt, T. Welton, Solvents and Green Chemistry, in Solvents and Solvent Effects in Organic Chemistry, fourth ed., Wiley-VCH Verlag, Weinheim, Germany, 2011, pp. 509–548. (Chapter 8).
[43] A. Mohammad, Inamuddin (Eds.), Green Solvents II: Properties and Applications of Ionic Liquids, Springer Verlag, Berlin and Heidelberg, Germany, 2012.
[44] V.V. Singh, A.K. Nigam, A. Batra, M. Boopathi, B. Singh, R. Vijayaraghavan, Applications of ionic liquids in electrochemical sensors and biosensors, Int. J. Electrochem., 2012 doi:10.1155/2012/165683. Article # 165683.
[45] Ionic Liquids, Themed Issue, Faraday Discuss. 154 (2012) 1.
[46] C.H. Arnaud, Ionic liquids improve separations, Chem. Eng. News (April 2, 2012) 36–37.
[47] Ionic Liquid, Wikipedia Article, 2012. http://en.wikipedia.org/wiki/Ionic_liquid.
[48] G.W. Meindersma, M. Maase, A.B. De Haan, Ionic liquids, seventh ed., Ullmann's Encyclopedia of Industrial Chemistry, vol. 19, Wiley-VCH, Weinheim, Germany, 2012, pp. 548–575.
[49] N.V. Plechkova, K.R. Seddon (Eds.), Ionic Liquids UnCOILed: Critical Expert Overviews, John Wiley & Sons, New York, NY, 2013.
[50] NASA Document JPR 5322.1, Contamination Control Requirements Manual, National Aeronautics and Space Administration, Johnson Space Center, Houston, TX, 2009.
[51] ESA Standard ECSS-Q-70-01B, Space Product Assurance – Cleanliness and Contamination Control, European Space Agency, Noordwijk, The Netherlands, 2008.
[52] IEST Standard IEST-STD-CC1246D, Product Cleanliness Levels and Contamination Control Program, Institute for Environmental Science and Technology (IEST), Rolling Meadows, IL, 2002.

[53] E.K. Stansbery, Genesis Discovery Mission Contamination Control Plan, NASA Document GN-460000–100 National Aeronautics and Space Administration, Johnson Space Center, Houston, TX, 1998.
[54] Sigma-Aldrich Corporation, St. Louis, MO. www.sigma-aldrich.com.
[55] Phosphonium Ionic Liquids, Cytec Industries, Woodland Park, NJ. www.cytec.com.
[56] Ionic Liquids, Covalent Associates, Corvallis, OR. www.covalentassociates.com.
[57] DuPont FluoroIntermediates. www.dupont.com/fluoroIntermediates.
[58] Proionic GmbH, Grambach, Austria. http://www.proionic.com/en/.
[59] Ionic Liquids Handbook. ACROS Organics, Geel, Belgium. www.acros.com.
[60] Ionic Liquids, Solvionic, Toulouse, France. www.solvionic.com.
[61] Ionic Liquids. IoLiTec Ionic Liquids Technologies GmbH, Heilbronn, Germany. www.iolitec.com.
[62] Ionic Liquids – Solutions for Your Success. BASF, Ludwigshafen, Germany. http://www.intermediates.basf.com/chemicals/ionic-liquids/index.
[63] Ionic Liquids, Solchemar, Lisbon, Portugal. www.solchemar.com.
[64] Ionic Liquids, Scionix Ltd, London, UK. www.scionix.co.uk.
[65] Kanto Chemical, Tokyo, Japan. www.kanto.co.jp.
[66] Imidazole Derivatives, Nippon Gohsei, Osaka, Japan. http://www.nichigo.co.jp/english/product/index02.html.
[67] Ionic Liquids, Chemada Fine Chemicals, HaNegev, Israel. www.chemada.com.
[68] A.D. McNaught, A. Wilkinson (Eds.), Compendium of Chemical Terminology, IUPAC Recommendations The "Gold Book", second ed., Blackwell Scientific Publications, Oxford, UK, 1997. Electronic interactive version http://goldbook.iupac.org/.
[69] J.H. Davis Jr., P.A. Fox, From curiosities to commodities: ionic liquids begin the transition, Chem. Commun. (2003) 1209–1212.
[70] S.A. Bolkan, J.T. Yoke, Room temperature fused salts based on copper(I) chloride-1-methyl-3-ethylimidazolium chloride mixtures. 1. Physical properties, J. Chem. Eng. Data 31 (1986) 194.
[71] S.P. Wicelinski, R.J. Gale, J.S. Wilkes, Differential scanning calorimetric study of low melting organic chlorogallate systems, Thermochim. Acta 26 (1988) 255.
[72] E.R. Schreiter, J.E. Stevens, M.F. Ortwerth, R.G. Freeman, A room-temperature molten salt prepared from $AuCl_3$ and 1-ethyl-3-methylimidazolium chloride, Inorg. Chem. 38 (1999) 3935.
[73] J.-Z. Yang, P. Tian, L.-L. He, W.-G. Xu, Studies on room temperature ionic liquid $InCl_3$-EMIC, Fluid Phase Equilib. 204 (2003) 295.
[74] J.-Z. Yang, P. Tian, W.-G. Xu, S.-Z. Liu, Studies on an ionic liquid prepared from $InCl_3$ and 1-methyl-3-butylimidazolium chloride, Thermochim. Acta 412 (2004) 1.
[75] S. Hayashi, H. Yamaguchi, Discovery of a magnetic ionic liquid [bmim]$FeCl_4$, Chem. Lett. 33 (2004) 1590.
[76] Y. Yoshida, J. Fujii, K. Muroi, A. Otsuka, G. Saito, M. Takahashi, T. Yoko, Highly conducting ionic liquids based on 1-ethyl-3-methylimidazolium cation, Synth. Met. 153 (2005) 421.
[77] P.G. Rickert, M.R. Antonio, M.A. Firestone, K.-A. Kubatko, T. Szreder, J.F. Wishart, M.L. Dietz, Tetraalkylphosphonium polyoxometalate ionic liquids: novel, organic–inorganic hybrid materials, J. Phys. Chem. B 111 (2007) 4685.
[78] S.-F. Tang, A. Babai, A.-V. Mudring, Low melting europium ionic liquids as luminescent soft materials, Angew. Chem. Int. Ed. 47 (2008) 7631.
[79] B. Mallick, B. Balke, C. Felser, A.-V. Mudring, Dysprosium room-temperature ionic liquids with strong luminescence and response to magnetic fields, Angew. Chem. Int. Ed. 47 (2008) 7635.

[80] Y. Yoshida, H. Tanaka, G. Saito, L. Ouahab, H. Yoshida, N. Sato, Valence-tautomeric ionic liquid composed of a cobalt bis(dioxolene) complex dianion, Inorg. Chem. 48 (2009) 9989.
[81] H.D. Pratt III, A.J. Rose, C.L. Staiger, D. Ingersoll, T.M. Anderson, Synthesis and characterization of ionic liquids containing copper, manganese, or zinc coordination cations, Dalton Trans. 40 (2011) 11396.
[82] F.M. Santos, P. Brandão, V. Félix, M.R.M. Domingues, J.S. Amaral, V.S. Amaral, H.I.S. Nogueira, A.M.V. Cavaleiro, Organic–inorganic hybrid materials based on iron(III)-polyoxotungstates and 1-butyl-3-methylimidazolium cations, Dalton Trans. 41 (2012) 12145.
[83] B. Mallick, A. Metlen, M. Nieuwenhuyzen, R.D. Rogers, A.-V. Mudring, Mercuric ionic liquids: [$C_n$mim][HgX$_3$], where n = 3, 4 and X = Cl, Br, Inorg. Chem. 51 (2012) 193.
[84] S. Hobby, Sandia National Laboratories Researchers Find Energy Storage "Solutions" in MetILs, Sandia Labs News Releases, Albuquerque, NM, February 17, 2012.
[85] T.L. Greaves, A. Weerawardena, C. Fong, I. Krodkiewska, C.J. Drummond, Protic ionic liquids: solvents with tunable phase behavior and physicochemical properties, J. Phys. Chem. B 110 (2006) 22479.
[86] K.N. Marsh, J.A. Boxall, R. Lichtenthaler, Room temperature ionic liquids and their mixtures, Fluid Phase Equilib. 219 (2004) 93.
[87] S. Gabriel, J. Weiner, Ueber einige Abkömmlinge des Propylamins, Ber. Dtsch. Chem. Ges. 21 (1888) 2669.
[88] Th. Curtius, Neues vom Stickstoffwasserstoff, Ber. Dtsch. Chem. Ges. 24 (1891) 3341.
[89] C. Schall, Über organische und geschmolzene Salze, Zeitschr. Elektrochem. 14 (1908) 397.
[90] P. Walden, Über die Molekulargrösse und elektrische Leitfähigkeit einiger geschmolzenen Salze, Bull. Acad. Sci. St. Petersburg (1914) 405–422.
[91] R.S. Kalb, M.J. Kotschan, Trioctylmethylammonium thiosalicylate (TOMATS). A novel, high-performance, task-specific ionic liquid for the extraction of heavy metals from aqueous solutions, Aldrich ChemFiles 6 (2006) 13.
[92] V.M. Egorov, D.I. Djigailo, D.S. Momotenko, D.V. Chernyshov, I.I. Torocheshnikova, S.V. Smirnova, I.V. Pletnev, Task-specific ionic liquid trioctylmethylammonium salicylate as extraction solvent for transition metal ions, Talanta 80 (2010) 1177.
[93] Z. Yacob, J. Liebscher, 1,2,3-Triazolium Salts as a Versatile New Class of Ionic Liquids, in: S.T. Handy (Ed.), Ionic Liquids – Classes and Properties, InTech Publishing – Open Access Company, Rijeka, Croatia, 2011, pp. 1–22. http://www.intechopen.com/books/ionic-liquids-classes-and-properties/1-2-3-triazolium-salts-as-a-versatile-new-class-of-ionic-liquids.
[94] S. Sanghi, E. Willett, C. Versek, M. Tuominen, E.B. Coughlin, Physicochemical properties of 1,2,3-triazolium ionic liquids, RSC Adv. 2 (2012) 848.
[95] S. Lago, H. Rodríguez, M.K. Khoshkbarchi, A. Soto, A. Arce, Enhanced oil recovery using the ionic liquid trihexyl(tetradecyl)phosphonium chloride: phase behaviour and properties, RSC Adv. 2 (2012) 9392.
[96] R. Kohli, Surface contamination removal using dense-phase fluids: liquid and supercritical carbon dioxide, in: R. Kohli, K.L. Mittal (Eds.), Developments in Surface Contamination and Cleaning, vol. 5, Elsevier, Oxford, UK, 2013, pp. 1–54. (Chapter 1).
[97] S. Keskin, D. Kayrak-Talay, U. Akman, O. Hortaçsu, A review of ionic liquids towards supercritical fluid applications, J. Supercrit. Fluids 43 (2007) 150.
[98] P.J. Carvalho, V.H. Álvarez, I.M. Marrucho, M. Aznar, J.A.P. Coutinho, High carbon dioxide solubilities in trihexyltetradecylphosphonium-based ionic liquids, J. Supercrit. Fluids 52 (2010) 258.
[99] Z. Sedláková, Z. Wagner, High pressure phase equilibria in systems containing CO_2 and ionic liquid of the [$C_n$mim][Tf$_2$N] type, Chem. Biochem. Eng. Q. 26 (2012) 55.

[100] S. Keskin, U. Akman, O. Hortaçsu, Continuous cleaning of contaminated soils using ionic liquids and supercritical CO_2, in: Proceedings of the Tenth European Meeting on Supercritical Fluids: Reactions, Materials and Natural Products, Colmar, France, 2005, Pi3.1–Pi3.6.

[101] S. Keskin, U. Akman, O. Hortaçsu, Soil remediation via an ionic liquid and supercritical CO_2, Chem. Eng. Proc. 47 (2008) 1693.

[102] A.A. Fannin Jr., D.A. Floreani, L.A. King, J.S. Landers, B.J. Piersma, D.J. Stech, R.L. Vaughn, J. Wilkes, J.L. Williams, Properties of 1,3-dialkylimidazolium chloride-aluminum chloride ionic liquids. 2. Phase transitions, densities, electrical conductivities, and viscosities, J. Phys. Chem. 88 (1984) 2614.

[103] C.L. Hussey, T.B. Scheffler, J.S. Wilkes, A.A. Fannin Jr., Chloroaluminate equilibria in the aluminum chloride–1–methyl–3–ethylimidazolium chloride ionic liquid, J. Electrochem. Soc. 133 (1986) 1389.

[104] C.M. Gordon, J.D. Holbrey, A.R. Kennedy, K.R. Seddon, Ionic liquid crystals: hexafluorophosphate salts, J. Mater. Chem. 8 (1998) 2627.

[105] J.D. Holbrey, K.R. Seddon, The phase behaviour of 1-alkyl-3-methylimidazolium tetrafluoroborates; ionic liquids and ionic crystals, J. Chem. Soc. Dalton Trans. (1999) 2133–2139.

[106] F. Neve, O. Francescangeli, A. Crispini, Crystal architecture and mesophase structure of long-chain n-alkylpyridinium tetrachlorometallates, Inorg. Chim. Acta 338 (2002) 51.

[107] J. Kärkkäinen, Preparation and characterization of some ionic liquids and their use in the dimerization reaction of 2-methylpropene, Ph.D. Dissertation, University of Oulu, Oulu, Finland, 2007.

[108] J.L. Solà Cervera, A. König, Recycling concept for aluminum electrodeposition from the ionic liquid system emim $[Tf_2N]$-$AlCl_3$, Chem. Eng. Technol. 33 (2010) 1979.

[109] D.C. Apperley, C. Hardacre, P. Licence, R.W. Murphy, N.V. Plechkova, K.R. Seddon, G. Sreenivasan, M. Swadźba-Kwaśny, I.J. Villar-Garcia, Speciation of chloroindate(III) ionic liquids, Dalton Trans. 39 (2010) 8679.

[110] M. Currie, J. Estager, P. Licence, S. Men, P. Nockemann, K.R. Seddon, M. Swadźba-Kwaśny, C. Terrade, Chlorostannate(II) ionic liquids: speciation, lewis acidity, and oxidative stability, Inorg. Chem. 52 (2013) 1710.

[111] J. Estager, P. Nockemann, K.R. Seddon, M. Swadźba-Kwaśny, S. Tyrrell, Validation of speciation techniques: a study of chlorozincate(II) ionic liquids, Inorg. Chem. 50 (2011) 5258.

[112] M.J. Earle, J.M.S.S. Esperança, M.A. Gilea, J.N. Canongia Lopes, L.P.N. Rebelo, J.W. Magee, K.R. Seddon, J.A. Widegren, The distillation and volatility of ionic liquids, Nature 439 (2006) 831.

[113] P. Wasserscheid, Chemistry: volatile times for ionic liquids, Nature 439 (2006) 797.

[114] J.P. Armstrong, C. Hurst, R.G. Jones, P. Licence, K.R.J. Lovelock, C.J. Satterley, I.J. Villar-Garcia, Vapourisation of ionic liquids, Phys. Chem. Chem. Phys. 9 (2007) 982.

[115] A.W. Taylor, K.R.J. Lovelock, A. Deyko, P. Licence, R.G. Jones, High vacuum distillation of ionic liquids and separation of ionic liquid mixtures, Phys. Chem. Chem. Phys. 12 (2010) 1772.

[116] M. Maase, Distillation of ionic liquids, U.S. Patent 7,754,053, 2010.

[117] K. Masonne, M. Siemer, W. Mormann, W. Leng, Distillation of ionic liquids, U.S. Patent Application 2010/0300870, 2010.

[118] J.M.S.S. Esperança, J.N.C. Lopes, M. Tariq, L.M.N.B.F. Santos, J.W. Magee, L.P.N. Rebelo, Volatility of aprotic ionic liquids—a review, J. Chem. Eng. Data 55 (2010) 3.

[119] K. Swiderski, A. McLean, C.M. Gordon, D.H. Vaughan, Estimates of internal energies of vaporisation of some room temperature ionic liquids, Chem. Commun. (2004) 2178–2179.

[120] S.H. Lee, S.B. Lee, The Hildebrand solubility parameters, cohesive energy densities and internal energies of 1-alkyl-3-methylimidazolium-based room temperature ionic liquids, Chem. Commun. (2005) 3469–3471.

[121] L.M.N.B.F. Santos, J.N.C. Lopes, J.A.P. Coutinho, J.M.S.S. Esperança, L.R. Gomes, I.M. Marrucho, L.P.N. Rebelo, Ionic liquids: first direct determination of their cohesive energy, J. Am. Chem. Soc. 129 (2007) 284.

[122] M.S. Kelkar, E.J. Maginn, Calculating the enthalpy of vaporization for ionic liquid clusters, J. Phys. Chem. B 111 (2007) 9424.

[123] A. Marciniak, The solubility parameters of ionic liquids, Int. J. Mol. Sci. 11 (2010) 1973.

[124] A. Marciniak, The Hildebrand solubility parameters of ionic liquids—part 2, Int. J. Mol. Sci. 12 (2011) 3553.

[125] H.A. Øye, M. Jagtoyen, T. Oksefjell, J.S. Wilkes, Vapour pressure and thermodynamics of the system 1-methyl-3-ethyl-imidazolium chloride – aluminium chloride, Mater. Sci. Forum 73–75 (1991) 183.

[126] Y.U. Paulechka, G.J. Kabo, A.V. Blokhin, O.A. Vydrov, J.W. Magee, M. Frenkel, Thermodynamic properties of 1-butyl-3-methylimidazolium hexafluorophosphate in the ideal gas state, J. Chem. Eng. Data 48 (2003) 457.

[127] Y.U. Paulechka, Dz. H. Zaitsau, G.J. Kabo, A.A. Strechan, Vapor pressure and thermal stability of ionic liquid 1-butyl-3-methylimidazolium bis(trifluoromethylsulfonyl)amide, Thermochim. Acta 439 (2005) 158.

[128] L.P.N. Rebelo, J.N. Canongia Lopes, J.M.S.S. Esperança, E. Filipe, On the critical temperature, normal boiling point, and vapor pressure of ionic liquids, J. Phys. Chem. B 109 (2005) 6040.

[129] Dz. H. Zaitsau, G.J. Kabo, A.A. Strechan, Y.U. Paulechka, A. Tschersich, S.P. Verevkin, A. Heintz, Experimental vapor pressures of 1-alkyl-3-methylimidazolium bis(trifluoromethylsulfonyl)imides and a correlation scheme for estimation of vaporization enthalpies of ionic liquids, J. Phys. Chem. A 110 (2006) 7303.

[130] J.A. Widegren, Y.-M. Wang, W.A. Henderson, J.W. Magee, Relative volatilities of ionic liquids by vacuum distillation of mixtures, J. Phys. Chem. B 111 (2007) 8959.

[131] V.N. Emel'yanenko, S.P. Verevkin, A. Heintz, The gaseous enthalpy of formation of the ionic liquid 1-butyl-3-methylimidazolium dicyanamide from combustion calorimetry, vapor pressure measurements, and ab initio calculations, J. Am. Chem. Soc. 129 (2007) 3930.

[132] H. Luo, G.A. Baker, S. Dai, Isothermogravimetric determination of the enthalpies of vaporization of 1-alkyl-3-methylimidazolium ionic liquids, J. Phys. Chem. B 112 (2008) 10077.

[133] O. Aschenbrenner, S. Supasitmongkol, M. Taylor, P. Styring, Measurement of vapour pressures of ionic liquids and other low vapour pressure solvents, Green Chem. 11 (2009) 1217.

[134] S.D. Chambreau, G.L. Vaghjiani, A. To, C. Koh, D. Strasser, O. Kostko, S.R. Leone, Heats of vaporization of room temperature ionic liquids by tunable vacuum ultraviolet photoionization, J. Phys. Chem. B 114 (2010) 1361.

[135] C. Wang, H. Luo, H. Li, S. Dai, Direct UV-spectroscopic measurement of selected ionic-liquid vapors, Phys. Chem. Chem. Phys. 12 (2010) 7246.

[136] M.A.A. Rocha, C.F.R.A.C. Lima, L.R. Gomes, B. Schröder, J.A.P. Coutinho, I.M. Marrucho, J.M.S.S. Esperanca, L.P.N. Rebelo, K. Shimizu, J.N.C. Lopes, L.M.N.B.F. Santos, High-accuracy vapor pressure data of the extended $[C_nC_1im][Ntf_2]$ ionic liquid series: trend changes and structural shifts, J. Phys. Chem. B 115 (2011) 10919.

[137] A. Deyko, K.R.J. Lovelock, J.-A. Corfield, A.W. Taylor, P.N. Gooden, I.J. Villar-Garcia, P. Licence, R.G. Jones, V.G. Krasovskiy, E.A. Chernikova, L.M. Kustov, Measuring and predicting $\Delta_{vap}H_{298}$ values of ionic liquids, Phys. Chem. Chem. Phys. 11 (2009) 8544.

[138] K.R.J. Lovelock, A. Deyko, P. Licence, R.G. Jones, Vaporisation of an ionic liquid near room temperature, Phys. Chem. Chem. Phys. 12 (2010) 8893.

[139] A. Deyko, S.G. Hessey, P. Licence, E.A. Chernikova, V.G. Krasovskiy, L.M. Kustov, R.G. Jones, The enthalpies of vaporisation of ionic liquids: new measurements and predictions, Phys. Chem. Chem. Phys. 14 (2012) 3181.

[140] Dz. H. Zaitsau, K. Fumino, V.N. Emel'yanenko, A.V. Yermalayeu, R. Ludwig, S.P. Verevkin, Structure–property relationships in ionic liquids: a study of the anion dependence in vaporization enthalpies of imidazolium-based ionic liquids, ChemPhysChem 13 (2012) 1868.

[141] S.P. Verevkin, Predicting the enthalpy of vaporization of ionic liquids: a simple rule for a complex property, Angew. Chem. Int. Ed. 47 (2008) 5071.

[142] M. Bier, S. Dietrich, Vapor pressure of ionic liquids, Mol. Phys. 108 (2010) 211.

[143] J.O. Valderrama, L.A. Forero, An analytical expression for the vapor pressure of ionic liquids based on an equation of state, Fluid Phase Equilib. 317 (2012) 77.

[144] E.J. Maginn, Molecular simulation of ionic liquids: current status and future opportunities, J. Phys.: Condens. Matter 21 (2009) 373101.

[145] E.J. Maginn, J.R. Elliott, Historical perspective and current outlook for molecular dynamics as a chemical engineering tool, Ind. Eng. Chem. Res. 49 (2010) 3059.

[146] F. Dommert, K. Wendler, R. Berger, L. Delle Site, Ch. Holm, Force fields for studying the structure and dynamics of ionic liquids: a critical review of recent developments, Chem. Phys. Chem. 13 (2012) 1625.

[147] S. Aparicio, M. Atilhan, F. Karadas, Thermophysical properties of pure ionic liquids: review of present situation, Ind. Eng. Chem. Res. 49 (2010) 9580.

[148] J.A.P. Coutinho, P.J. Carvalho, N.M.C. Oliveira, Predictive methods for the estimation of thermophysical properties of ionic liquids, RSC Adv. 2 (2012) 7322.

[149] E.I. Izgorodina, Theoretical Approaches to Ionic Liquids: From Past History to Future Directions, in: N.V. Plechkova, K.R. Seddon (Eds.), Ionic Liquids UnCOILed: Critical Expert Oveviews, John Wiley & Sons, New York, NY, 2013, pp. 181–230. (Chapter 6).

[150] M. Diedenhofen, A. Klamt, K.N. Marsh, A. Schäfer, Prediction of the vapor pressure and vaporization enthalpy of 1-n-alkyl-3-methylimidazolium-bis-(trifluoromethanesulfonyl) amide ionic liquids, Phys. Chem. Chem. Phys. 9 (2007) 4653.

[151] C.M. Hansen, Hansen Solubility Parameters: A User's Handbook, second ed., CRC Press, Boca Raton, FL, 2007.

[152] D. Camper, P. Scovazzo, C. Koval, R. Noble, Gas solubilities in room-temperature ionic liquids, Ind. Eng. Chem. Res. 43 (2004) 3049.

[153] H. Jin, B. O'Hare, J. Dong, S. Arzhantsev, G.A. Baker, J.F. Wishart, A.J. Benesi, M. Maroncelli, Physical properties of ionic liquids consisting of the 1-butyl-3-ethylimidazolium cation with various anions and the bis(trifluoromethylsulfonyl)imide anion with various cations, J. Phys. Chem. B 112 (2008) 81.

[154] P.K. Kilaru, P. Scovazzo, Correlations of low-pressure carbon dioxide and hydrocarbon solubilities in imidazolium-, phosphonium-, and ammonium-based room-temperature ionic liquids. Part 2. Using activation energy of viscosity, Ind. Eng. Chem. Res. 47 (2008) 910.

[155] S.S. Moganty, R.E. Baltus, Regular solution theory for low pressure carbon dioxide solubility in room temperature ionic liquids: ionic liquid solubility parameter from activation energy of viscosity, Ind. Eng. Chem. Res. 49 (2010) 5846.

[156] M.J. Kamlet, J.-L.M. Abboud, M.H. Abraham, R.W. Taft, Linear solvation energy relationships. 23. A comprehensive collection of the solvatochromic parameters, π^*, α, and β, and some methods for simplifying the generalized solvatochromic equation, J. Org. Chem. 48 (1983) 2877.

[157] P. Scovazzo, D. Camper, J. Kieft, J. Poshusta, C. Koval, R. Noble, Regular solution theory and CO_2 gas solubility in room-temperature ionic liquids, Ind. Eng. Chem. Res. 43 (2004) 6855.

[158] D. Camper, C. Becker, C. Koval, R.D. Noble, Low pressure hydrocarbon solubility in room temperature ionic liquids containing imidazolium rings interpreted using regular solution theory, Ind. Eng. Chem. Res. 44 (2005) 1928.

[159] A. Finotello, J.E. Bara, D. Camper, R.D. Noble, Room-temperature ionic liquids: temperature dependence of gas solubility selectivity, Ind. Eng. Chem. Res. 47 (2008) 3453.

[160] J. Gupta, C. Nunes, S. Vyas, S. Jonnalagadda, Prediction of solubility parameters and miscibility of pharmaceutical compounds by molecular dynamics simulations, J. Phys. Chem. B 115 (2011) 2014.

[161] K. Paduszyński, J. Chiyen, D. Ramjugernath, T.M. Letcher, U. Domańska, Liquid–liquid phase equilibrium of (piperidinium-based ionic liquid + an alcohol) binary systems and modelling with NRHB and PCP-SAFT, Fluid Phase Equilib. 305 (2011) 43.

[162] M.L.S. Batista, C.M.S.S. Neves, P.J. Carvalho, R. Gani, J.A.P. Coutinho, Chameleonic behavior of ionic liquids and its impact on the estimation of solubility parameters, J. Phys. Chem. B 115 (2011) 12879.

[163] S.P. Verevkin, V.N. Emel'yanenko, Dz. H. Zaitsau, R.V. Ralys, Ionic liquids: differential scanning calorimetry as a new indirect method for determination of vaporization enthalpies, J. Phys. Chem. B 116 (2012) 4276.

[164] T. Ueki, M. Watanabe, Polymers in ionic liquids: dawn of neoteric solvents and innovative materials, Bull. Chem. Soc. Jpn. 85 (2012) 33.

[165] Y.S. Sistla, L. Jain, A. Khanna, Validation and prediction of solubility parameters of ionic liquids for CO_2 capture, Sep. Purif. Technol. 97 (2012) 51.

[166] K. Paduszyński, U. Domańska, Thermodynamic modeling of ionic liquid systems: development and detailed overview of novel methodology based on the PC-SAFT, J. Phys. Chem. B 116 (2012) 5002.

[167] B. Iliev, M. Smiglak, A. Świerczyńska, T.J.S. Schubert, Miscibility and phase separation of various ionic liquids with common organic solvents and water, ILSEPT First International Conference on Ionic Liquids in Separation and Purification Technology, Stiges, Spain, 2011. http://www.iolitec.de/en/Download-document/669-2011-ILSEPT-Solubility.html.

[168] M.D. Bermejo, A. Martín, Solubility of gases in ionic liquids, Global J. Phys. Chem. 2 (2011) 324.

[169] X. Ji, C. Held, G. Sadowski, Modeling imidazolium-based ionic liquids with ePC-SAFT, Fluid Phase Equilib. 335 (2012) 64.

[170] R. Ludwig, Thermodynamic properties of ionic liquids – a cluster approach, Phys. Chem. Chem. Phys. 10 (2008) 4333.

[171] M. Roth, Partitioning behaviour of organic compounds between ionic liquids and supercritical fluids, J. Chromatogr. A 1216 (2009) 1861.

[172] L.F. Vega, O. Vilaseca, F. Llovell, J.S. Andreu, Modeling ionic liquids and the solubility of gases in them: recent advances and perspectives, Fluid Phase Equilib. 294 (2010) 15.

[173] A.R. Ferreira, M.G. Freire, J.C. Ribeiro, F.M. Lopes, J.G. Crespo, J.A.P. Coutinho, An overview of the liquid–liquid equilibria of (ionic liquid + hydrocarbon) binary systems and their modeling by the conductor-like screening model for real solvents, Ind. Eng. Chem. Res. 50 (2011) 5279.

[174] G. Annat, M. Forsyth, D.R. MacFarlane, Ionic liquid mixtures – variations in physical properties and their origins in molecular structure, J. Phys. Chem. B 116 (2012) 8251.

[175] V.S. Bernales, A.V. Marenich, R. Contreras, C.J. Cramer, D.G. Truhlar, Quantum mechanical continuum solvation models for ionic liquids, J. Phys. Chem. B 116 (2012) 9122.

[176] F. Llovell, E. Valente, O. Vilaseca, L.F. Vega, Modeling complex associating mixtures with [C_n-mim][Tf_2N] ionic liquids: predictions from the soft-SAFT equation, J. Phys. Chem. B 115 (2011) 4387.

[177] M.B. Oliveira, F. Llovell, J.A.P. Coutinho, L.F. Vega, Modeling the [NTf_2] pyridinium ionic liquids family and their mixtures with the soft statistical associating fluid theory equation of state, J. Phys. Chem. B 116 (2012) 9089.

[178] F. Llovell, M. Belo, O. Vilaseca, J.A.P. Coutinho, L.F. Vega, Thermodynamic modeling of the solubility of supercritical CO_2 and other gases on ionic liquids with the soft-SAFT equation of state, in: Proceedings of the Tenth International Symposium on Supercritical Fluids, ISSF, 2012, J. Supercrit. Fluids (2013). http://issf2012.com/handouts/documents/333_004.pdf.

[179] L.A. Blanchard, D. Hancu, E.J. Beckman, J.F. Brennecke, Green processing using ionic liquids and CO_2, Nature 399 (1999) 28.

[180] J.F. Brennecke, E.J. Maginn, Ionic liquids: innovative fluids for chemical processing, AIChE J. 47 (2001) 2384.

[181] L.A. Blanchard, Z. Gu, J.F. Brennecke, High-pressure phase behavior of ionic liquid/CO_2 systems, J. Phys. Chem. B 105 (2001) 2437.

[182] A.M. Scurto, S.N.V.K. Aki, J.F. Brennecke, CO_2 as a separation switch for ionic liquid/organic mixtures, J. Am. Chem. Soc. 124 (2002) 10276.

[183] C. Cadena, J.L. Anthony, J.K. Shah, T.I. Morrow, J.F. Brennecke, E.J. Maginn, Why is CO_2 so soluble in imidazolium-based ionic liquids? J. Am. Chem. Soc. 126 (2004) 5300.

[184] A. Shariati, C.J. Peters, High pressure phase equilibria of systems with ionic liquids, J. Supercrit. Fluids 34 (2005) 171.

[185] A. Adamou, J.-J. Letourneau, E. Rodier, R. David, P. Guiraud, Characterization of the ionic liquid bmimPF_6 in supercritical conditions, in: Proceedings of the Tenth European Meeting on Supercritical Fluids: Reactions, Materials and Natural Products, Colmar, France, 2005 Pi5.1-Pi5.6.

[186] K.I. Gutkowski, A. Shariati, B. Breure, S.B. Bottini, E.A. Brignole, C.J. Peters, Experiments and modelling of systems with ionic liquids, in: Proceedings of the Tenth European Meeting on Supercritical Fluids: Reactions, Materials and Natural Products, Colmar, France, 2005 Pi1.1-Pi1.6.

[187] K.I. Gutkowski, A. Shariati, C.J. Peters, High-pressure phase behavior of the binary ionic liquid system 1-octyl-3-methylimidazolium tetrafluoroborate + carbon dioxide, J. Supercrit. Fluids 39 (2006) 187.

[188] A. Shariati, S. Raeissi, C.J. Peters, CO_2 Solubility in Alkylimidazolium-Based Ionic Liquids, in: T.M. Letcher (Ed.), Developments and Applications in Solubility, RSC Publishing, Cambridge, UK, 2007, pp. 131–152.

[189] S. Mattedi, P.J. Carvalho, J.A.P. Coutinho, V.H. Alvarez, M. Iglesias, High pressure CO_2 solubility in n-methyl-2-hydroxyethylammonium protic ionic liquids, J. Supercrit. Fluids 56 (2011) 224.

[190] Pure Component Properties, Queriable Database. CHERIC Chemical Engineering Research Information Center, Seoul, South Korea, 2012. http://www.cheric.org/research/kdb/hcprop/cmpsrch.php.

[191] F.J.V. Santos, C.A.N. de Castro, J.H. Dymond, N.K. Dalaouti, M.J. Assael, A. Nagashima, Standard reference data for the viscosity of toluene, J. Phys. Chem. Ref. Data 35 (2006) 1.

[192] Cyclohexane Data Sheet, Sunoco Chemicals, Philadelphia, PA, 2012.

[193] J.A. Widegren, J.W. Magee, Density, viscosity, speed of sound, and electrolytic conductivity for the ionic liquid 1-hexyl-3-methylimidazolium bis(trifluoromethylsulfonyl)imide and its mixtures with water, J. Chem. Eng. Data 52 (2007) 2331.

[194] K.N. Marsh, J.F. Brennecke, R.D. Chirico, M. Frenkel, A. Heintz, J.W. Magee, C.J. Peters, L.P.N. Rebelo, K.R. Seddon, Thermodynamic and thermophysical properties of the reference ionic liquid: 1-hexyl-3-methylimidazolium bis[(trifluoromethyl)sulfonyl]amide (including mixtures). Part 1. Experimental methods and results (IUPAC technical report), Pure Appl. Chem. 81 (2009) 781.
[195] R.D. Chirico, V. Diky, J.W. Magee, M. Frenkel, K.N. Marsh, Thermodynamic and thermophysical properties of the reference ionic liquid: 1-hexyl-3-methylimidazolium bis[(trifluoromethyl)sulfonyl]amide (including mixtures). Part 2. Critical evaluation and recommended property values (IUPAC technical report), Pure Appl. Chem. 81 (2009) 791.
[196] G.L. Burrell, I.M. Burgar, F. Separovic, N.F. Dunlop, Preparation of protic ionic liquids with minimal water content and [16]N NMR study of proton transfer, Phys. Chem. Chem. Phys. 12 (2010) 1571.
[197] P. Hapiot, C. Lagrost, Electrochemical reactivity in room-temperature ionic liquids, Chem. Rev. 108 (2008) 2238.
[198] O. Borodin, G.D. Smith, H. Kim, Viscosity of a room temperature ionic liquid: predictions from nonequilibrium and equilibrium molecular dynamics simulations, J. Phys. Chem. B 113 (2009) 4771.
[199] F. Castiglione, G. Raos, G.B. Appetecchi, M. Montanino, S. Passerini, M. Moreno, A. Famulari, A. Mele, Blending ionic liquids: how physico-chemical properties change, Phys. Chem. Chem. Phys. 12 (2010) 1784.
[200] C.A.N. de Castro, Thermophysical properties of ionic liquids: do we know how to measure them accurately? J. Mol. Liq. 156 (2010) 10.
[201] J.N.C. Lopes, M.F. Costa Gomes, P. Husson, A.A.H. Pdua, L.P.N. Rebelo, S. Sarraute, M. Tariq, Polarity, viscosity, and ionic conductivity of liquid mixtures containing [C_4C_1im][Ntf_2] and a molecular component, J. Phys. Chem. B 115 (2011) 6988.
[202] G. Yu, D. Zhao, L. Wen, S. Yang, X. Chen, Viscosity of ionic liquids: database, observation, and quantitative structure–property relationship analysis, AIChE J. 58 (2012) 2885.
[203] S.N. Butler, F. Müller-Plathe, A molecular dynamics study of viscosity in ionic liquids directed by quantitative structure–property relationships, ChemPhysChem. 13 (2012) 1791.
[204] J.C.F. Diogo, F.J.P. Caetano, J.M.N.A. Fareleira, W.A. Wakeham, C.A.M. Afonso, C.S. Marques, Viscosity measurements of the ionic liquid trihexyl(tetradecyl)phosphonium dicyanamide [$P_{6,6,6,14}$][dca] using the vibrating wire technique, J. Chem. Eng. Data 57 (2012) 1015.
[205] K.R. Seddon, A. Stark, M.J. Torres, Influence of chloride, water, and organic solvents on the physical properties of ionic liquids, Pure Appl. Chem. 72 (2000) 2275.
[206] J.G. Huddleston, A.E. Visser, W.M. Reichert, H.D. Willauer, G.A. Broker, R.D. Rogers, Characterization and comparison of hydrophilic and hydrophobic room temperature ionic liquids incorporating the imidazolium cation, Green Chem. 3 (2001) 156.
[207] C. Villagrán, M. Deetlefs, W.R. Pitner, C. Hardacre, Quantification of halide in ionic liquids using ion chromatography, Anal. Chem. 76 (2004) 2118.
[208] J.A. Widegren, A. Laesecke, J.W. Magee, The effect of dissolved water on the viscosities of hydrophobic room-temperature ionic liquids, Chem. Commun. (2005) 1610–1612.
[209] D.S. Silvester, R.G. Compton, Electrochemistry in room temperature ionic liquids: a review and some possible applications, Z. Phys. Chem. 220 (2006) 1247.
[210] A.K. Burrell, R.E. Del Sesto, S.N. Baker, T.M. McCleskey, G.A. Baker, The large scale synthesis of pure imidazolium and pyrrolidinium ionic liquids, Green Chem. 9 (2007) 449.
[211] K.R. Harris, M. Kanakubo, L.A. Woolf, Temperature and pressure dependence of the viscosity of the ionic liquids 1-hexyl-3-methylimidazolium hexafluorophosphate and

1-butyl-3-methylimidazolium bis(trifluoromethylsulfonyl)imide, J. Chem. Eng. Data 52 (2007) 1080.
[212] S. Randström, M. Montanino, G.B. Appetecchi, C. Lagergren, A. Moreno, S. Passerini, Effect of water and oxygen traces on the cathodic stability of n-alkyl-n-methylpyrrolidinium bis(trifluoromethanesulfonyl)imide, Electrochim. Acta 53 (2008) 6397.
[213] P.J. Carvalho, T. Regueira, L.M.N.B.F. Santos, J. Fernandez, J.A.P. Coutinho, Effect of water on the viscosities and densities of 1-butyl-3-methylimidazolium dicyanamide and 1-butyl-3-methylimidazolium tricyanomethane at atmospheric pressure, J. Chem. Eng. Data 55 (2010) 645.
[214] H.V. Spohr, G.N. Patey, The influence of water on the structural and transport properties of model ionic liquids, J. Chem. Phys. 132 (2010) 234510.
[215] M.A. Firestone, J.A. Dzielawa, P. Zapol, L.A. Curtiss, S. Seifert, M.L. Dietz, Lyotropic liquid–crystalline gel formation in a room-temperature ionic liquid, Langmuir 18 (2002) 7258.
[216] O. Green, S. Grubjesic, S. Lee, M.A. Firestone, The design of polymeric ionic liquids for the preparation of functional materials, Polym. Rev. 49 (2009) 339.
[217] M. Yoshizawa, W. Xu, C.A. Angell, Ionic liquids by proton transfer: vapor pressure, conductivity, and the relevance of ΔpK_a from aqueous solutions, J. Am. Chem. Soc. 125 (2003) 15411.
[218] W. Xu, E.I. Cooper, C.A. Angell, Ionic liquids: ion mobilities, glass temperatures, and fragilities, J. Phys. Chem. B 107 (2003) 6170.
[219] H. Tokuda, S. Tsuzuki, M.A.H. Susan, K. Hayamizu, M. Watanabe, How ionic are room-temperature ionic liquids? An indicator of the physicochemical properties, J. Phys. Chem. B 110 (2006) 19593.
[220] R. Hayes, G.G. Warr, R. Atkin, At the interface: solvation and designing ionic liquids, Phys. Chem. Chem. Phys. 12 (2010) 1709.
[221] K. Ueno, H. Tokuda, M. Watanabe, Ionicity in ionic liquids: correlation with ionic structure and physicochemical properties, Phys. Chem. Chem. Phys. 12 (2010) 1649.
[222] C. Schreiner, S. Zugmann, R. Hartl, H.J. Gores, Fractional Walden rule for ionic liquids: examples from recent measurements and a critique of the so-called ideal KCl line for the Walden plot, J. Chem. Eng. Data 55 (2010) 1784.
[223] D.R. MacFarlane, M. Forsyth, E.I. Izgorodina, A.P. Abbott, G. Annata, K. Fraser, On the concept of ionicity in ionic liquids, Phys. Chem. Chem. Phys. 11 (2009) 4962.
[224] H. Liu, E. Maginn, An MD study of the applicability of the Walden rule and the Nernst–Einstein model for ionic liquids, Chem. Phys. Chem. 13 (2012) 1701.
[225] R. Kohli, Methods for monitoring and measuring cleanliness of surfaces, in: R. Kohli, K.L. Mittal (Eds.), Developments in Surface Contamination and Cleaning, vol. 4, Elsevier, Oxford, UK, 2012, pp. 107–178. (Chapter 3).
[226] S. Kuwabata, A. Kongkanand, D. Oyamatsu, T. Torimoto, Observation of ionic liquid by scanning electron microscope, Chem. Lett. 35 (2006) 600.
[227] S. Arimoto, M. Sugimura, H. Kageyama, T. Torimoto, S. Kuwabata, Development of new techniques for scanning electron microscope observation using ionic liquid, Electrochim. Acta 53 (2008) 6228.
[228] S. Kuwabata, T. Tsuda, T. Torimoto, Room-temperature ionic liquid as new medium for material production and analyses under vacuum conditions, J. Phys. Chem. Lett. 1 (2010) 3177.
[229] Y. Ishigaki, Y. Nakamura, T. Takehara, N. Nemoto, T. Kurihara, H. Koga, H. Nakagawa, T. Takegami, N. Tomosugi, S. Miyazawa, S. Kuwabata, Ionic liquid enables simple and rapid sample preparation of human culturing cells for scanning electron microscope analysis, Microsc. Res. Tech. 74 (2011) 415.

[230] Y. Ishigaki, Y. Nakamura, T. Takehara, T. Kurihara, H. Koga, T. Takegami, H. Nakagawa, N. Nemoto, N. Tomosugi, S. Kuwabata, S. Miyazawa, Comparative study of hydrophilic and hydrophobic ionic liquids for observing cultured human cells by scanning electron microscopy, Microsc. Res. Tech. 74 (2011) 1104.
[231] T. Tsuda, N. Nemoto, K. Kawakami, E. Mochizuki, S. Kishida, T. Tajiri, T. Kushibiki, S. Kuwabata, SEM observation of wet biological specimens pretreated with room-temperature ionic liquid, Chembiochem. 12 (2011) 2547.
[232] A. Dwiranti, L. Lin, E. Mochizuki, S. Kuwabata, A. Takaoka, S. Uchiyama, K. Fukui, Chromosome observation by scanning electron microscopy using ionic liquid, Microsc. Res. Tech. 75 (2012) 1113.
[233] T. Torimoto, K. Okazaki, T. Kiyama, K. Hirahara, N. Tanaka, S. Kuwabata, Sputter deposition onto ionic liquids: simple and clean synthesis of highly dispersed ultrafine metal nanoparticles, Appl. Phys. Lett. 89 (2006) 243117.
[234] K. Okazaki, T. Kiyama, K. Hirahara, N. Tanaka, S. Kuwabata, T. Torimoto, Single-step synthesis of gold–silver alloy nanoparticles in ionic liquids by a sputter deposition technique, Chem. Commun. (2008) 691–693.
[235] K. Ueno, K. Hata, T. Katakabe, M. Kondoh, M. Watanabe, Nanocomposite ion gels based on silica nanoparticles and an ionic liquid: ionic transport, viscoelastic properties, and microstructure, J. Phys. Chem. B 112 (2008) 9013.
[236] K. Ueno, A. Inaba, M. Kondoh, M. Watanabe, Colloidal stability of bare and polymer-grafted silica nanoparticles in ionic liquids, Langmuir 24 (2008) 5253.
[237] K. Ueno, A. Inaba, Y. Sano, M. Kondoh, M. Watanabe, A soft glassy colloidal array in ionic liquid, which exhibits homogeneous, non-brilliant and angle-independent structural colours, Chem. Commun. (2009) 3603–3605.
[238] D. Yoshimura, T. Yokoyama, T. Nishi, H. Ishii, R. Ozawa, H. Hamaguchi, K. Seki, Electronic structure of ionic liquids at the surface studied by UV photoemission, J. Electron Spectrosc. Relat. Phenom. 144–147 (2005) 319.
[239] E.F. Smith, F.J.M. Rutten, I.J. Villar-Garcia, D. Briggs, P. Licence, Ionic liquids in vacuo: analysis of liquid surfaces using ultra-high-vacuum techniques, Langmuir 22 (2006) 9386.
[240] O. Höfft, S. Bahr, M. Himmerlich, S. Krischok, J.A. Schaefer, V. Kempter, Electronic structure of the surface of the ionic liquid [EMIM][Tf$_2$N] studied by metastable impact electron spectroscopy (MIES), UPS, and XPS, Langmuir 22 (2006) 7120.
[241] C. Aliaga, C.S. Santos, S. Baldelli, Surface chemistry of room-temperature ionic liquids, Phys. Chem. Chem. Phys. 9 (2007) 3683.
[242] K. Nakajima, A. Ohno, M. Suzuki, K. Kimura, Observation of molecular ordering at the surface of trimethylpropylammonium bis(trifluoromethanesulfonyl)imide using high-resolution Rutherford backscattering spectroscopy, Langmuir 24 (2008) 2282.
[243] K.R.J. Lovelock, C. Kolbeck, T. Cremer, N. Paape, P.S. Schulz, P. Wasserscheid, F. Maier, H.-P. Steinrück, Influence of different substituents on the surface composition of ionic liquids studied using ARXPS, J. Phys. Chem. B 113 (2009) 2854.
[244] F.J.M. Rutten, H. Tadesse, P. Licence, Rewritable imaging on the surface of frozen ionic liquids, Angew. Chem. Int. Ed. 46 (2007) 4163.
[245] P.J. Dyson, M.A. Henderson, J.S. McIndoe, Ionic liquids: solutions for electrospray ionisation mass spectrometry, in: N.V. Plechkova, R.D. Rogers, K.R. Seddon (Eds.), Ionic Liquids: From Knowledge to Application, ACS Symposium Series, vol. 1030, American Chemical Society, Washington, D.C., 2009, pp. 135–146.
[246] K. Kanai, T. Nishi, T. Iwahashi, Y. Ouchi, K. Seki, Y. Harada, S. Shin, Electronic structures of imidazolium-based ionic liquids, J. Electron Spectrosc. Relat. Phenom. 174 (2009) 110.

[247] M. Holzweber, E. Pittenauer, H. Hutter, Investigation of ionic liquids under Bi-ion and Bi-cluster ions bombardment by ToF-SIMS, J. Mass Spectrom. 45 (2010) 1104.
[248] Y. Fujiwara, N. Saito, H. Nonaka, A. Suzuki, T. Nakanaga, T. Fujimoto, A. Kurokawa, S. Ichimura, Time-of-flight secondary ion mass spectrometry (TOF-SIMS) using the metal-cluster-complex primary ion of $Ir_4(CO)_7^+$, Surf. Interface Anal. 43 (2011) 245.
[249] I. Niedermaier, C. Kolbeck, N. Taccardi, P.S. Schulz, J. Li, T. Drewello, P. Wasserscheid, H.-P. Steinrück, F. Maier, Organic reactions in ionic liquids studied by in situ XPS, ChemPhysChem. 13 (2012) 1725.
[250] S. Bovio, A. Podestà, P. Milani, Investigation of Interfacial Properties of Supported [C_4mim][NTf_2] Thin Films by Atomic Force Microscopy, in: N.V. Plechkova, R.D. Rogers, K.R. Seddon (Eds.), Ionic Liquids: From Knowledge to Application, ACS Symposium Series, vol. 1030, American Chemical Society, Washington, D.C., 2009, pp. 273–290.
[251] R. Atkin, G.G. Warr, Bulk and interfacial nanostructure in protic room temperature ionic liquids, in: N.V. Plechkova, R.D. Rogers, K.R. Seddon (Eds.), Ionic Liquids: From Knowledge to Application, ACS Symposium Series, vol. 1030, American Chemical Society, Washington, D.C., 2009, pp. 317–333.
[252] R. Atkin, S.Z. El Abedin, R. Hayes, L.H.S. Gasparotto, N. Borisenko, F. Endres, AFM and STM studies on the surface interaction of [BMP]TFSA and [EMIM]TFSA ionic liquids with Au(111), J. Phys. Chem. C 113 (2009) 13266.
[253] Y. Yokota, T. Harada, K. Fukui, Direct observation of layered structures at ionic liquid/solid interfaces by using frequency-modulation atomic force microscopy, Chem. Commun. 46 (2010) 8627.
[254] A. Labuda, P. Grütter, Atomic force microscopy in viscous ionic liquids, Langmuir 28 (2012) 5319.
[255] T. Ichii, M. Fujimura, M. Negami, K. Murase, H. Sugimura, Frequency modulation atomic force microscopy in ionic liquid using quartz tuning fork sensors, Jpn. J. Appl. Phys. 51 (2012) 08KB08.
[256] N. Borisenko, S.Z. El Abedin, F. Endres, An in-situ STM and DTS study of the extremely pure [EMIM]FAP/Au(111) interface, Chemphyschem. 13 (2012) 1736.
[257] C. Pereira, I. Ferreira, L.C. Branco, I.C.A. Sandu, T. Busani, Atomic Force Microscopy as a Valuable Tool in an Innovative Multi-Scale and Multi-Technique Non-Invasive Approach to Surface Cleaning Monitoring, Youth in the Conservation of Cultural Heritage, YOCOCU 2012, Antwerpen, Belgium, June 2012. www.yococu.com/6_Paintings/A-08.doc.
[258] J.L. Anthony, E.J. Maginn, J.F. Brennecke, Solution thermodynamics of imidazolium-based ionic liquids and water, J. Phys. Chem. B 105 (2001) 10942.
[259] D.S.H. Wong, J.P. Chen, J.M. Chang, C.H. Chou, Phase equilibria of water and ionic liquids [emim][PF_6] and [bmim][PF_6], Fluid Phase Equilib. 194–197 (2002) 1089.
[260] J. McFarlane, W.B. Ridenour, H. Luo, R.D. Hunt, D.W. DePaoli, R.X. Ren, Room temperature ionic liquids for separating organics from produced water, Sep. Sci. Technol. 40 (2005) 1245.
[261] D.J. Couling, R.J. Bernot, K.M. Docherty, J.K. Dixon, E.J. Maginn, Assessing the factors responsible for ionic liquid toxicity to aquatic organisms via quantitative structure–property relationship modeling, Green Chem. 8 (2006) 82.
[262] M.G. Freire, L.M.N.B.F. Santos, A.M. Fernandes, J.A.P. Coutinho, I.M. Marrucho, An overview of the mutual solubilities of water–imidazolium-based ionic liquids systems, Fluid Phase Equilib. 261 (2007) 449.
[263] T. Kakiuchi, Mutual solubility of hydrophobic ionic liquids and water in liquid–liquid two-phase systems for analytical chemistry, Anal. Sci. 24 (2008) 1221.

[264] J. Ranke, A. Othman, P. Fan, A. Müller, Explaining ionic liquid water solubility in terms of cation and anion hydrophobicity, Int. J. Mol. Sci. 10 (2009) 1271.
[265] K. Řehák, P. Morávek, M. Strejc, Determination of mutual solubilities of ionic liquids and water, Fluid Phase Equilib. 316 (2012) 17.
[266] T. Zhou, L. Chen, Y. Ye, L. Chen, Z. Qi, H. Freund, K. Sundmacher, An overview of mutual solubility of ionic liquids and water predicted by COSMO-RS, Ind. Eng. Chem. Res. 51 (2012) 6256.
[267] Integrated Laboratory Systems, Inc., Ionic Liquids 1-butyl-3-methylimidazolium chloride (CAS No. 79917-90-1), 1-butyl-1-methylpyrrolidinium chloride (CAS No. 479500-35-1) n-butylpyridinium chloride (CAS No. 1124-64-7). Review of Toxicological Literature, Report for National Toxicology Program (NTP)/National Institute of Environmental Health Sciences (NIEHS), Research Triangle Park, NC, 2004. http://ntp.niehs.nih.gov/ntp/htdocs/Chem_Background/ExSumPdf/Ionic_liquids.pdf.
[268] P.J. Scammells, J.L. Scott, R.D. Singer, Ionic liquids: the neglected issues, Aust. J. Chem. 58 (2005) 155.
[269] K.M. Docherty, C.F. Kulpa Jr., Toxicity and Antimicrobial Activity of Imidazolium and Pyridinium Ionic Liquids, Green Chem. 7 (2005) 185.
[270] D. Zhao, Y. Liao, Z. Zhang, Toxicity of ionic liquids, CLEAN: Soil, Air, Water 35 (2007) 42.
[271] J. Ranke, A. Müller, U. Bottin-Weber, F. Stock, S. Stolte, J. Arning, R. Störmann, B. Jastorff, Lipophilicity parameters for ionic liquid cations and their correlation to in vitro cytotoxicity, Ecotoxicol. Environ. Saf. 67 (2007) 430.
[272] J. Ranke, S. Stolte, R. Störmann, J. Arning, B. Jastorff, Design of sustainable chemical products – the example of ionic liquids, Chem. Rev. 107 (2007) 2183.
[273] T.P.T. Pham, C.W. Cho, Y.S. Yun, Environmental fate and toxicity of ionic liquids: a review, Water Res. 44 (2010) 352.
[274] R.F.M. Frade, C.A.M. Afonso, Impact of ionic liquids in environment and humans: an overview, Hum. Exp. Toxicol. 29 (2010) 1038.
[275] D. Coleman, N. Gathergood, Biodegradation studies of ionic liquids, Chem. Soc. Rev. 39 (2010) 600.
[276] N. Wood, Toxicity of Ionic Liquids and Organic Solvents towards *Escherichia coli* and *Pseudomonas putida*, Ph.D. Dissertation, The University of Manchester, Manchester, UK, 2011.
[277] E. Liwarska-Bizukojc, D. Gendaszewska, Removal of imidazolium ionic liquids by microbial associations: study of the biodegradability and kinetics, J. Biosci. Bioeng. 115 (2013) 71.
[278] The UFT/Merck Ionic Liquids Biological Effects Database, University of Bremen, Bremen, Germany, 2012. http://www.il-eco.uft.uni-bremen.de.
[279] J. Arning, S. Stolte, A. Böschen, F. Stock, W.-R. Pitner, U. Welz-Biermann, B. Jastorff, J. Ranke, Qualitative and quantitative structure activity relationships for the inhibitory effects of cationic head groups, functionalised side chains and anions of ionic liquids on acetylcholinesterase, Green Chem. 10 (2008) 47.
[280] A. García-Lorenzo, E. Tojo, J. Tojo, M. Teijeira, F.J. Rodríguez-Berrocal, M.P. González, V.S. Martínez-Zorzano, Cytotoxicity of selected imidazolium-derived ionic liquids in the human caco-2 cell line. Sub-structural toxicological interpretation through a QSAR study, Green Chem. 10 (2008) 508.
[281] J.S. Torrecilla, J. García, E. Rojo, F. Rodríguez, Estimation of toxicity of ionic liquids in *leukemia rat cell line* and *acetylcholinesterase* enzyme by principal component analysis, neural networks and multiple lineal regressions, J. Hazard. Mater. 164 (2009) 182.
[282] M. Alvarez-Guerra, A. Irabien, Design of ionic liquids: an ecotoxicity (*Vibrio fischeri*) discrimination approach, Green Chem. 13 (2011) 1507.

[283] N. Wood, G. Stephens, Accelerating the discovery of biocompatible ionic liquids, Phys. Chem. Chem. Phys. 12 (2010) 1670.
[284] S. Zhang, N. Sun, X. He, X. Lu, X. Zhang, Physical properties of ionic liquids: database and evaluation, J. Phys. Chem. Ref. Data 35 (2006) 1475.
[285] S. Zhang, X. Lu, Q. Zhou, X. Li, X. Zhang, S. Li (Eds.), Ionic Liquids: Physicochemical Properties, Elsevier, Oxford, UK, 2009.
[286] A. Stark, J.-A. van den Berg, The Handbook of Ionic Liquids: Data, Wiley-Interscience, New York, NY, 2011.
[287] Reaxys Overview. Reed Elsevier Properties SA, Amsterdam, The Netherlands, 2012. www.reaxys.com.
[288] SciFinder, Chemical Abstracts Service, American Chemical Society, Columbus, OH, 2012. www.cas.org/products/scifinder.
[289] Beilstein Handbuch der organischen Chemie. Leopold Voss Verlag, Hamburg and Leipzig, Germany, 1998.
[290] Gmelin Handbook of Inorganic and Organometallic Chemistry. Springer Verlag, Berlin and Heidelberg, Germany, 1997.
[291] IL THERMO Ionic Liquids Database, NIST Standard Reference Database #147 National Institute of Science and Technology, Boulder, CO, 2012. http://ilthermo.boulder.nist.gov/ILThermo/mainmenu.uix.
[292] DDBSP 2011 – Ionic Liquids, Dortmund Data Bank Software & Separation Technology GmbH, Oldenburg, Germany, 2012. www.ddbst.com.
[293] Delph-IL Ionic Liquids Database. Novionic Inc, Dailin, China, 2012. http://delphil.net/web/html/features.html.
[294] Deep Eutectic Solvent, Wikipedia Article, 2012. http://en.wikipedia.org/w/index.php?oldid=493816010.
[295] A.P. Abbott, G. Capper, D.L. Davies, R.K. Rasheed, V. Tambyrajah, Novel solvent properties of choline chloride/urea mixtures, Chem. Commun. (2003) 70–71.
[296] A.P. Abbott, D. Boothby, G. Capper, D.L. Davies, R.K. Rasheed, Deep eutectic solvents formed between choline chloride and carboxylic acids: versatile alternatives to ionic liquids, J. Am. Chem. Soc. 126 (2004) 9142.
[297] A.P. Abbott, G. Capper, D.L. Davies, K.J. McKenzie, S.U. Obi, Solubility of metal oxides in deep eutectic solvents based on choline chloride, J. Chem. Eng. Data 51 (2006) 1280.
[298] A.P. Abbott, J.C. Barron, K.S. Ryder, D. Wilson, Eutectic-based ionic liquids with metal-containing anions and cations, Chem. Eur. J. 13 (2007) 6495.
[299] H.G. Morrison, C.C. Sun, S. Neervannan, Characterization of thermal behavior of deep eutectic solvents and their potential as drug solubilization vehicles, Int. J. Pharm. 378 (2009) 136.
[300] K. Haerens, E. Matthijs, A. Chmielarz, B. Van der Bruggen, The use of ionic liquids based on choline chloride for metal deposition: a green alternative? J. Environ. Manage. 90 (2009) 3245.
[301] M.A. Kareem, F.S. Mjalli, M. Ali Hashim, I.M. Al-Nashef, Phosphonium-based ionic liquids analogues and their physical properties, J. Chem. Eng. Data 55 (2010) 4632.
[302] I.M. Al Nashef, S.M. Al Zahrani, Process for the destruction of halogenated hydrocarbons and their homologous/analogous in deep eutectic solvents at ambient conditions, U.S. Patent No. 7,812,211, 2010.
[303] I.M. Al Nashef, S.M. Al Zahrani, Method for the preparation of reactive hydrogen peroxide in deep eutectic solvents, U.S. Patent No. 7,763,768, 2010.
[304] O. Ciocirlan, O. Iulian, O. Croitoru, Effect of temperature on the physico-chemical properties of three ionic liquids containing choline chloride, Rev. Chim. (Bucharest) 61 (2010) 721.

[305] R.F. Miller, Deep eutectic solvents and applications, U.S. Patent No. 8,022,014, 2011.
[306] M. Hall, P. Bansal, J.H. Lee, M.J. Realff, A.S. Bommarius, Biological pretreatment of cellulose: enhancing enzymatic hydrolysis rate using cellulose-binding domains from cellulases, Bioresour. Technol. 102 (2011) 2910.
[307] Q. Zhang, K. De Oliveira Vigier, S. Royer, F. Jérôme, Deep eutectic solvents: syntheses, properties and applications, Chem. Soc. Rev. 41 (2012) 7108.
[308] D.P.R. Thanu, S. Raghavan, Benign Deep Eutectic Solvents for Replacement of Organic Solvents Based Cleaning Formulations in BEOL Cleaning, TECHCON Proceedings, Semiconductor Research Corporation, 2010. pp. 1–4.
[309] D.P.R. Thanu, S. Raghavan, M. Keswani, Use of urea-choline chloride eutectic solvent for back end of line cleaning applications, Electrochem. Solid-State Lett. 14 (2011) H358.
[310] D.P.R. Thanu, Use of Dilute Hydrofluoric Acid and Deep Eutectic Solvent Systems for Back End of Line Cleaning in Integrated Circuit Fabrication, Ph.D. Dissertation, University of Arizona, Tucson, AZ, 2011.
[311] D.P.R. Thanu, S. Raghavan, M. Keswani, Effect of water addition to choline chloride-urea deep eutectic solvent (DES) on the removal of post-etch residues formed on copper, IEEE Trans. Semicond. Manuf. 25 (2012) 516.
[312] J. Taubert, M. Keswani, S. Raghavan, Post-etch residue removal using choline chloride–malonic acid deep eutectic solvent (DES), Microelectron. Eng. 102 (2013) 81.
[313] E.B. Borghi, S.P. Ali, P.J. Morando, M.A. Blesa, Cleaning of stainless steel surfaces and oxide dissolution by malonic and oxalic acids, J. Nucl. Mater. 229 (1996) 115.
[314] D. García, V.I.E. Bruyère, R. Bordoni, A.M. Olmedo, P.J. Morando, Malonic acid: a potential reagent in decontamination processes for Ni-rich alloy surfaces, J. Nucl. Mater. 412 (2011) 72.
[315] J.B. Durkee, Will ionic liquids be useful cleaning chemicals? Controlled Environ. Mag. 11 (January 2008) 29.
[316] P.E. Rakita, Challenges to the Commercial Production of Ionic Liquids, in: R.D. Rogers, K.R. Seddon (Eds.), Ionic Liquids as Green Solvents, ACS Symposium Series, vol. 856, American Chemical Society, Washington, D.C., 2003, pp. 32–40. (Chapter 3).
[317] S.S. Seelig, A. O'Lenick, Green solvents and ionic liquids: formulating for the sustainable future, in: Proceedings of the 101st American Oil Chemists' Society Annual Meeting and Exposition, Phoenix, AZ, 2010. http://www.aocs.org/files/ampresentation/35844_fulltext.pdf.
[318] V.R. Koch, C. Nanjundiah, R.T. Carlin, Hydrophobic ionic liquids, U.S. Patent 5,827,602, 1998.
[319] B. Mertens, M. Mondin, N. Andries, J. Massaux, All purpose liquid cleaning composition comprising anionic, amine oxide and EO-BO nonionic surfactant, U.S. Patent 6,020,296, 2000.
[320] A.P. Abbott, D.L. Davies, G. Capper, R.K. Rasheed, V. Tambyrajah, Ionic liquids and their use as solvents, WIPO Patent WO 2002/026701, 2002.
[321] A.B. McEwen, Cyclic delocalized cations connected by spacer groups, U.S. Patent 6,513,241, 2003.
[322] K.N. Price, R.T. Hartshorn, R.H. Rohrbaugh, W.M. Scheper, M.S. Showell, K.H. Baker, M.R. Sivik, J.J. Scheibel, R.R. Gardner, P.K. Reddy, J.D. Aiken, M.C. Addison, Ionic liquid based products and method of using the same, U. S. Patent Application 20060240728, 2006.
[323] K. Binnemans, A.C. Görller-Walrand, P. Nockemann, B. Thijs, Novel ionic liquids, WIPO Patent WO 2007/147222, 2007.
[324] R. Bacardit, P. Giordani, M. Rigamonti, D. Bankmann, M. Del Mar Combarros Gracia, Ionic liquid composition for the removal of oxide scale, WIPO Patent Application WO 2010/052123, 2010.

[325] S.E. Hecht, S.L. Cron, J.J. Scheibel, G.S. Miracle, K.R. Seddon, M. Earle, H.Q.N. Gunaratne, Ionic liquids derived from surfactants, U.S. Patent Application 2010/0099314, 2010.
[326] S.E. Hecht, G.S. Miracle, S.L. Cron, M.S. Showell, Ionic liquids derived from peracid anions, U.S. Patent 7,786,065, 2010.
[327] S.E. Hecht, K.N. Price, P.S. Berger, P.R. Foley, H.D. Hutton, III, M.S. Showell, R.R. Gardner, R.L. Niehoff, K.R. Seddon, H.Q.N. Gunaratne, M.J. Earle, Multiphase cleaning compositions having ionic liquid phase, U.S. Patent 7,928,053, 2011.
[328] R.F. Miller, Deep eutectic solvents and their applications, U.S. Patent 8,022,014, 2011.
[329] Y. Chauvin, L. Magna, G.P. Niccolai, J.-M. Basset, Imidazolium salts and the use of these ionic liquids as a solvent, U.S. Patent 7,915,426, 2011.
[330] S. Dai, H. Luo, Synthesis of ionic liquids, U.S. Patent 8,049,026, 2011.
[331] R.J. Small, Semiconductor cleaning using superacids, U.S. Patent 7,923,424, 2011.
[332] T. Beyersdorff, T. Schubert, Ionic liquids as antistatic additives in surface cleaning processes, in: Proceedings of the 44th International Detergency Conference, Düsseldorf, Germany, 2009.
[333] A. Bösmann, T. Schubert, Identification of Industrial Applications for Ionic Liquids: High-Performance-Additives for the Use in Hi-Tech-Cleaning-Solutions, Poster IoLiTec Ionic Liquids Technologies GmbH, Heilbronn, Germany, 2012. www.iolitec.de/en/Poster/Page-4.html.
[334] T. Beyersdorff, T. Schubert, Ionic Liquids as Antistatic Additives in Cleaning Solutions – The Wandres Process, IoLiTec Ionic Liquids Technologies GmbH, Heilbronn, Germany, 2012. www.aails.com/is_ia_wp.asp.
[335] Ingomat-cleaner CF 05 – Micro-Cleaning of Flat Surfaces, Product Information, Wandres Micro-Cleaning GmbH, Buchenbach, Germany, 2012. www.wandres.com.
[336] P. Schwab, S. Kempka, M. Seiler, B. Gloeckler, Performance additives for improving the wetting properties of ionic liquids on solid surfaces, U.S. Patent Application 2010/0029519, 2010.
[337] H. Herzog, P. Schwab, M. Naumann, Process for producing antistatically treated artificial stone for flat structures, U.S. Patent Application 2010/0192814, 2010.
[338] ITB Inc, Precision Cleaning of Oxygen Systems and Components, Final Report NASA/CR-2009–214757, 2008.
[339] ITB Inc, Precision Cleaning of Oxygen Systems and Components. Phase II – Oxygen Cleaning Products Preliminary Testing, Final Report NASA DO27 FR O2Clean RM 09 25 09 F 2009.
[340] F. Malbosc, M. Bevon, S. Fantini, H. Olivier-Bourbigou, Procédé de nettoyage de surfaces mettant en oeuvre un liquide ionique de type protique, WIPO Patent WO 2010/040917, 2009.
[341] J. Yerty, M. Klingenberg, E. Berman, N. Voevodin, Are ionic liquids right for your parts cleaning Job? Prod. Finish. Mag. 76 (April 2012) 38.
[342] J. Yerty, M. Klingenberg, E. Berman, N. Voevodin, Ionic liquids for cleaning operations at Air Force logistics centers: part II, Prod. Finish. Mag. 76 (October 2012) 38. http://www.pfonline.com/articles/ionic-liquids-for-cleaning-operations-at-air-force-air-logistics-centers-part-ii.
[343] I. Ivanov, B. Soklev, S. Karakashev, Cleaning properties of ionic liquids, Second Workshop on Size-Dependent Effects in Materials for Environmental Protection and Energy application, SIZEMAT2, Nessebar, Bulgaria, 2010. http://sizemat2.igic.bas.bg/abstracts/SizeMat2_Abstract_Ivan_Ivanov_Topic_B.pdf.
[344] J.W. Krumpfer, P. Bian, P. Zheng, L. Gao, T.J. McCarthy, Contact angle hysteresis on superhydrophobic surfaces: an ionic liquid probe fluid offers mechanistic insight, Langmuir 27 (2011) 2166.

[345] K. Shukri, New Ionic Liquid Solvent Technology to Transform Metal Finishing. Products and Processes (IONMET), IONMET Report, Deutsche Gesellschaft für Galvano- und Oberflächentechnik e.V., Hilden, Germany, 2007. www.ionmet.org.
[346] A.P. Abbott, K.S. Ryder, U. König, Electrofinishing of metals using eutectic based ionic liquids, Trans. Inst. Met. Finish. 86 (2008) 196.
[347] J. Collins, Electropolishing Using Ionic Liquids, C-Tech Innovation Ltd, Chester, UK, 2008. http://www.prosurf-online.eu/fileadmin/documents/training/20080221_Birmingham/8_CTECH_JC_Ionmet_polishing.pdf.
[348] V. Cherginets, Oxide Solubilities in Ionic Melts, in: G. Wypych (Ed.), Handbook of Solvents, ChemTec Publishing, Toronto, Canada, 2001, pp. 1484–1496. (Chapter 21.3).
[349] P. Schwab, Use of ionic liquids as an additive for cleaning processes and/or supercritical gas, U.S. Patent Application 2010/0016205, 2010.
[350] P. Painter, P. Williams, E. Mannebach, A. Lupinsky, Systems, methods and compositions for the separation and recovery of hydrocarbons from particulate matter, U.S. Patent Application 2011/0042318, 2011.
[351] P. Painter, P. Williams, E. Mannebach, A. Lupinsky, Analogue ionic liquids for the separation and recovery of hydrocarbons from particulate matter, U.S. Patent Application 2012/0048783, 2012.
[352] Wastes – Hazardous Waste – Treatment, Storage & Disposal (TSD). U.S. Environmental Protection Agency, Washington, D.C., 2012. www.epa.gov/osw/hazard/tsd/index.htm.
[353] I.M. Al Nashef, S.M. Al Zahrani, Method for the preparation of reactive hydrogen peroxide in deep eutectic solvents, U.S. Patent 7,763,768, 2010.
[354] I.M. Al Nashef, S.M. Al Zahrani, Process for the destruction of halogenated hydrocarbons and their homologous/analogous in deep eutectic solvents at ambient conditions, U.S. Patent 7,812,211, 2010.
[355] W.M. Nelson, Ionic Liquid as Solvent, Catalyst Support: Chemical Agent Decontamination and Detoxification, Report 45571.1-CH, U.S. Army Research Office, Research Triangle Park, NC, 2004.
[356] O. Zech, A. Harrar, W. Kunz, Nonaqueous Microemulsions Containing Ionic Liquids – Properties and Applications, in: A. Kokorin (Ed.), Ionic Liquids: Theory, Properties, New Approaches, InTech Publishing – Open Access Company, Rijeka, Croatia, 2011978-953-307-349-1, pp. 245–270. www.intechopen.com/books/ionic-liquids-theory-properties-new-approaches. (Chapter 11).
[357] L. Carson, P.K.W. Chau, M.J. Earle, M.A. Gilea, B.F. Gilmore, S.P. Gorman, M.T. McCann, K.R. Seddon, Antibiofilm activities of 1-alkyl-3-methylimidazolium chloride ionic liquids, Green Chem. 11 (2009) 492.
[358] W.L. Hough-Troutman, M. Smiglak, S. Griffin, W.M. Reichert, I. Mirska, J. Jodynis-Liebert, T. Adamska, J. Nawrot, M. Stasiewicz, R.D. Rogers, J. PernakIonic, Ionic liquids with dual biological function: sweet and anti-microbial, hydrophobic quaternary ammonium-based salts, New J. Chem. 33 (2009) 26.
[359] A. Busetti, D.E. Crawford, M.J. Earle, M.A. Gilea, B.F. Gilmore, S.P. Gorman, G. Laverty, A.F. Lowry, M. McLaughlin, K.R. Seddon, Antimicrobial and antibiofilm activities of 1-alkylquinolinium bromide ionic liquids, Green Chem. 12 (2010) 420.
[360] M.H. Ismail, M. El-Harbawi, Y.A. Noaman, M.A. Bustam, N.B. Alitheen, N.A. Affandi, G. Hefter, C.Y. Yin, Synthesis and anti-microbial activity of hydroxylammonium ionic liquids, Chemosphere 84 (2011) 101.
[361] B.F. Gilmore, M.J. Earle, Development of ionic liquid biocides against microbial biofilms. Designer microbicides for infection control, Chem. Today 29 (2011) 50.

[362] B.F. Gilmore, Antimicrobial Ionic Liquids, in: A. Kokorin (Ed.), Ionic Liquids: Applications and Perspectives, InTech Publishing – Open Access Company, Rijeka, Croatia, 2011978-953-307-248-7, pp. 587–603. (Chapter 26).
[363] B.S. Sekhon, Ionic liquids: pharmaceutical and biotechnological applications, Asian J. Pharm. Biol. Res. 1 (2011) 395.
[364] A. Walker, Ionic Liquids for Natural Product Extraction, White paper, Bioniqs Ltd, Heslington, York, UK, 2008. http://www.slideserve.com/swaantje/ionic-liquids-for-natural-product-extraction-adam-walker-bioniqs-ltd.
[365] N. Carmona, M.A. Villegas, J.M. Fernández Navarro, Characterisation of an intermediate decay phenomenon of historical glasses, J. Mater. Sci. 41 (2006) 2339.
[366] A. Machado, P. Redol, L. Branco, M. Vilarigues, Medieval stained glass cleaning with ionic liquids, in: Proceedings of the IIC 2010 Congress – Conservation and the Eastern Mediterranean, Istanbul, Turkey, 2010. http://www.iiconservation.org/node/2769/c2010machado.pdf.
[367] A. Machado, P. Redol, L. Branco, M. Vilarigues, Ionic liquids for medieval stained glass cleaning: a new frontier, ICOM-CC Lisbon 2011: Sixteenth Triennial Conference, International Council of Museums – Committee for Conservation, Preprints CD, ISBN: 978-989-97522-0-7, 2011.
[368] C. Pereira, I.C.A. Sandu, L. Branco, T. Busani, I. Ferreira, Innovative Multi-Scale and Multi-Technique Approach to the Study of Enzyme-Based Cleaning of Varnish Layers, Second International Workshop on Physical and Chemical Analytical Techniques in Cultural Heritage, Lisbon, Portugal, 2012. http://cheri.cii.fc.ul.pt/BOOK_Abstracts_2nd_WORKSHOP_P&CATCH.pdf.
[369] B. von Gilsa, Gemäldereinigung mit Enzymen, Harzseifen und Emulsionen, Zeit. Kunsttechnologie Konservierung 5 (1991) 48.
[370] C. Todaro, Gil enzimi: limiti e potenzialità nel campo della pulitura delle pitture murali, in: G. Biscontin, G. Driussi (Eds.), XXI International Congress Scienza e Beni Culturali: Sulle Pitture Murali. Riflessione, Cconoscenze, Interventi, Arcadia Ricerche, Venice, Italy, 2005, pp. 487–496. http://www.arcadiaricerche.it/editoria/2005.htm.
[371] B.J. Palmer, D. Fu, R. Card, M.J. Miller, Scale removal, U.S. Patent 6,924,253, 2005.
[372] R. Kalb, H. Hofstätter, Method of treating a borehole and drilling fluid, U.S. Patent Application 2012/0103614, 2012.
[373] R.P. Swatloski, S.K. Spear, J.D. Holbrey, R.D. Rogers, Dissolution of cellulose with ionic liquids, J. Am. Chem. Soc. 124 (2002) 4974.
[374] H. Wang, G. Gurau, R.D. Rogers, Ionic liquid processing of cellulose, Chem. Soc. Rev. 41 (2012) 1519.
[375] J. Grävsik, D.G. Raut, J.-P. Mikkola, Challenges and Perspectives of Ionic Liquids vs. Traditional Solvents for Cellulose Processing, in: J. Mun, H. Sim (Eds.), Handbook of Ionic Liquids: Properties, Applications and Hazards, Nova Science Publishers, Hauppauge, NY, 2012978-1-62100-477-6, pp. 1–34. (Chapter 1).
[376] Processing Cellulose with Ionic Liquids. White Paper, BASF SE, Ludwigshafen, Germany, 2012. http://www.intermediates.basf.com/chemicals/web/en/content/products-and-industries/ionic-liquids/applications/cellulose_processing.
[377] P.S. Kulkarni, C.A.M. Afonso, Deep desulfurization of diesel fuel using ionic liquids: current status and future challenges, Green Chem. 12 (2010) 1139.
[378] R. Martínez-Palou, P.F. Sánchez, Perspectives of Ionic Liquids Applications for Clean Oilfield Technologies, in: A. Kokorin (Ed.), Ionic Liquids: Theory, Properties, New Approaches, InTech Publishing – Open Access Company, Rijeka, Croatia, 2011978-953-307-349-1, pp. 567–630. www.intechopen.com/books/ionic-liquids-theory-properties-new-approaches. (Chapter 24).

[379] P. Davey, Application study: ionic liquids in consumer products, Perfum. and Flavor. Mag., April 2008. http://www.perfumerflavorist.com/fragrance/application/industrial/16761536.html.
[380] A.J. O'Lenick Jr., Comparatively speaking: simple salt vs. ionic liquid, Cosmet. Toiletries Mag., November 2010. http://www.cosmeticsandtoiletries.com/research/techtransfer/108669424.html.
[381] Gem Sparkle Ionic Cleaning Liquid, Kassoy LLC, Plainview, NY, 2012. www.kassoy.com/gem-sparkle-ionic-cleaning-liquid.html.

Chapter 2

Microemulsions for Cleaning Applications

Lirio Quintero
Baker Hughes, Houston, TX, USA

Norman F. Carnahan
Carnahan Corporation, Houston, TX, USA

Chapter Outline

1. Introduction   66
2. Types of Microemulsions, Formulations and Properties   66
    2.1. Formulations   67
       2.1.1. Selection of Surfactant   70
       2.1.2. Effect of Salinity   70
       2.1.3. Effect of Cosurfactant   71
       2.1.4. Effect of Linkers   71
       2.1.5. Type of Oil or Solvent   73
       2.1.6. Temperature Effect   74
    2.2. Properties of Microemulsions   74
       2.2.1. Solubilization   74
       2.2.2. Interfacial Tension   75
       2.2.3. Contact Angle and Wettability   76
3. Basic Process and Principles of Cleaning Surfaces   77
4. Surface Cleaning and Contaminant Removal with Microemulsions   81
5. Design of Microemulsion Cleaners and Evaluation Techniques   81
6. Microemulsion Cleaning Applications   82
    6.1. Cleaning of Oil-Contaminated Drill Cuttings   84
    6.2. Wellbore Cleanup During Displacement of Oil-Based Drilling Fluids to Water-Based Fluid   85
    6.3. Near-Wellbore Cleaning in Oil and Gas Wells   88
       6.3.1. Removal of Oil-Based Fluid Filter Cake in OH Completion Wells   89
       6.3.2. Removal of Formation Damage in CH Completion Wells   92
    6.4. Other Cleaning Applications   93
       6.4.1. Wastewater Cleaning and Microemulsion Froth Flotation   94
       6.4.2. Microemulsion Cleaning of Contaminated Soil and Groundwater   94

6.4.3.	Microemulsion Cleaning of Textiles	96	
6.4.4.	Microemulsion Cleaning Using Nonaqueous Solvents	97	
6.4.5.	Microemulsion Cleaning of Building Exteriors	97	
6.4.6.	Microemulsion Cleaning of Frescoes and Artwork	97	
6.4.7.	Microemulsion Cleaning of Crude Oil Reservoirs	98	
6.4.8.	Microemulsion Cleaning of Fracturing Gels from Shale and Other Rock Formations	99	
7.	Current Trends and Future Developments	100	
	References	101	

1. INTRODUCTION

Since microemulsions were first described by Schulman et al. [1,2], the interest of researchers in studying their phase behavior, formulations, properties, fundamental mechanisms and applications has progressively grown. Salager reported that an exponential growth in microemulsion publications surpassed 1000 in 2003 [3]. This number keeps growing, with a considerable number of publications on microemulsion applications in recent years. Projecting the Salager-Schulman estimate, the number of microemulsion publications would be about 3200 in 2013, and around 7000 in 2020.

Microemulsions are thermodynamically stable, translucent fluids consisting of microdomains of oil and/or water stabilized by an interfacial film of surfactant molecules [4–8]. These systems may include optional additives such as cosurfactants, acids, lipophilic and hydrophilic linkers. Due to the importance of microemulsions in cleaning applications for various industries, their phase behavior and properties have been and continue to be studied extensively.

The ultralow interfacial tensions (IFTs) between oil and aqueous phases and the solubilization characteristics encountered in microemulsion systems make them useful for a wide variety of applications. Microemulsions are used in applications such as cleaning of surfaces, detergent applications, improved crude oil recovery, liquid–liquid extractions, pharmaceuticals and cosmetic formulations.

This chapter presents a comprehensive discussion on microemulsions for surface cleaning.

2. TYPES OF MICROEMULSIONS, FORMULATIONS AND PROPERTIES

Microemulsions are classified into three categories according to Winsor's phase behavior studies [9]. Winsor I microemulsion systems consist of oil-swollen micelles in a water phase in equilibrium with excess oil. Winsor II microemulsion systems consist of water-swollen reverse micelles in an oil phase in equilibrium with excess water. Winsor III systems are a middle-phase bicontinuous microemulsion in equilibrium with excess water and oil. The surfactant(s) in the bicontinuous microemulsions have equal affinity for the water phase and the oil phase. A single-phase microemulsion (Winsor IV) is obtained when a sufficient amount of surfactant is added to a Winsor III system to solubilize the excess oil and water into the microemulsion [4,5].

2.1. Formulations

The success of microemulsions as cleaners depends on the proper selection of additives in formulating the microemulsion fluid. Selection of the best fluid formulations requires systematic studies of phase behavior of the microemulsion systems. These studies include variables such as the type and concentration of surfactant, cosurfactant, salt, lipophilic linker, hydrophilic linker and solvent or oil, as well as temperature. Pressure has minimal or no effect except in cases where one of the components is compressible, e.g. systems involving supercritical fluids.

The phase behavior of water–surfactant–oil systems is typically studied by preparing a series of vials in which only one variable is progressively changed. For example, in a phase behavior study as a function of the concentration of sodium chloride (NaCl) salt, the percentage of NaCl is increased in each vial and the proportions of the remaining components of the system (surfactant, cosurfactant, aqueous phase, and oil) are maintained constant.

Figure 2.1 shows an example of Winsor phase behavior that could be obtained for oil–water–surfactant systems when the aforementioned variables are systematically changed. The phase behavior shows the characteristic progression from two-phase to three-phase to two-phase coexistence of an oil–water–surfactant system with the oil/water ratio, surfactant and cosurfactant

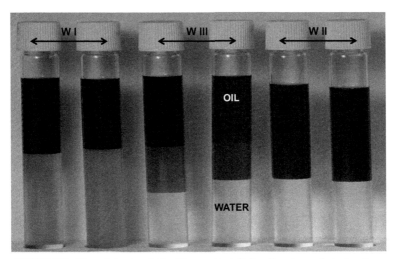

FIGURE 2.1 Phase behavior obtained by varying salinity oil–water–surfactant system.

concentrations constant and progressive increase of the salinity of the aqueous phase. At low salinity, a microemulsion coexists with an excess of oil (Winsor I). At high salinity, a microemulsion coexists with an excess of brine (Winsor II). At intermediate salinities, a bicontinuous microemulsion coexists with an excess of water phase and oil phase [9–12].

The data obtained from phase behavior studies are used to build the phase diagrams, enabling better understanding of capabilities and possible performance of the fluid formulated with a particular brine–surfactant–oil system.

Winsor introduced the concept of ratio of interactions (R) between the surfactant, oil, and water phases to determine the convexity of the interface and the resulting phase behavior [13].

$$R = (A_{SO} - A_{OO} - A_{TT})/(A_{SW} - A_{WW} - A_{HH}) \quad (2.1)$$

The term A_{SO} represents the interaction energy between the surfactant and the oil. A_{SW} is the interaction energy between the surfactant and the aqueous phase, A_{OO}, the interaction energy between oil molecules, A_{WW}, the interaction energy between water molecules, A_{TT}, the interaction energy between the tails of the surfactant molecules, and A_{HH}, the interaction energy between the surfactant heads.

If the surfactant molecules at the interface have stronger interactions with the water phase than the oil phase, A_{SW} is greater than A_{SO} and the term R is less than one. In this case, the interface shows a curvature toward the oil. The reverse effect (where R is greater than one) is observed when A_{SO} is greater than A_{SW}. In the case of equal interaction, R is equal to one. This means the interfacial curvature is near zero, which corresponds to the bicontinuous microemulsion structure described by various authors. Figure 2.2 shows an example of phase diagrams obtained with the phase behavior data of a brine–surfactant–oil system, as the ratio of interactions changes from $R < 1$ to $R > 1$.

There are various empirical correlations used to explain the physical chemistry of the formulations prepared in a phase behavior study. Salager et al. [14]

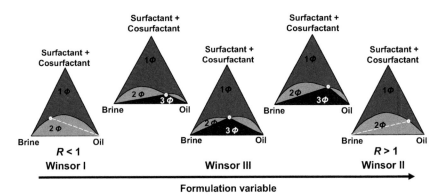

FIGURE 2.2 Phase diagrams of brine oil–water–surfactant blend system.

described the experimental technique to identify the optimal formulation in a phase behavior or formulation scan study. The optimum formulation in a series of vials of the phase behavior evaluation corresponds to the value of the variable with the three phases (Winsor III) that show equal volumes of oil and water solubilized. Equations (2.2) and (2.3) show the empirical correlation for the optimum formulation obtained for anionic surfactants and nonionic surfactants, respectively [14–20].

$$\ln S - K \times ACN - f(A) + \sigma - a_T(T - T_{ref}) = 0 \qquad (2.2)$$

$$\alpha = EON + b \times S - k \times ACN - \varphi(A) + C_T(T - T_{ref}) = 0 \qquad (2.3)$$

where,

S is the optimum salinity of the microemulsion system, expressed in wt% NaCl with respect to the aqueous phase

K is a constant for a given surfactant that depends on the type of surfactant head group

ACN is the linear alkane carbon number of the oil (for a nonlinear hydrocarbon oil or for a nonhydrocarbon oil it becomes the equivalent alkane carbon number, EACN)

$f(A)$ and $\phi(A)$ are functions of the alcohol/cosurfactant type and concentration

σ and α are the parameters that are functions of surfactant structure

a_T is a constant (~0.01 when temperature is in Celsius)

T is the temperature of evaluation

T_{ref} is a reference temperature

EON is the average number of ethylene oxide group per molecule of nonionic surfactant

b, k, a_T and c_T are the empirical constants that depend on the type of system.

These empirical correlations are actually numerical expressions of the surfactant affinity difference (SAD) used to interpret the formulation parameters of microemulsions [5,21,22]. The SAD expression represents the difference between the chemical potential of the surfactant in the aqueous phase and the oil phase as a function of different formulation parameters, as follows in Eqns (2.4) and (2.5) for anionic surfactants and for nonionic surfactants, respectively.

$$SAD/RT = -\ln S + k \times ACN + f(A) - \sigma + a_T \times (T - T_{ref}) \qquad (2.4)$$

$$SAD/RT = \alpha - EON + b \times S - k \times ACN - \phi(A) + c_T \times (T - T_{ref}) \qquad (2.5)$$

SAD can be negative, zero or positive. The optimum formulation is found where SAD is equal to zero. When the formulation has a deviation from the SAD = 0 condition, it can be modified to reach the optimum formulation by changing the physicochemical conditions (change the variables). Similar to the Winsor R ratio, the magnitude of the SAD value measures the departure from the optimum formulation. SAD < 1, SAD = 0 or SAD > 1 suggest the formation of Winsor Type I, Type III or Type II microemulsions, respectively.

Salager et al. proposed an empirical correlation known as the hydrophilic–lipophilic deviation (HLD) as a dimensionless form of the thermodynamically derived SAD equation to describe microemulsion systems [23,24]. In this case, negative-, zero-, or positive-HLD values suggest the formation of Winsor Type I, Type III or Type II microemulsions, respectively.

The general HLD equations for anionic and nonionic surfactants are Eqns (2.6) and (2.7), respectively [20,23]:

$$\text{HLD} = \ln S - k \times \text{ACN} - f(A) + \sigma - a_T \times (\Delta T) \quad (2.6)$$

$$\text{HLD} = \text{SAD}/RT = \alpha - \text{EON} + b \times S - k \times (\text{ACN}) - \phi(A) + c_T \times (\Delta T) \quad (2.7)$$

HLD is equal to zero at the optimum formulation because there is no difference between the hydrophilic and lipophilic interaction energies.

2.1.1. Selection of Surfactant

The selection of surfactant(s) for microemulsions depends on the requirements of specific applications. Microemulsions for cleaning applications (where variables such as temperature, type and density of brine, type of surface, and type of dirt or contaminant to be cleaned are considered in the design) usually require combinations of surfactants. The selection includes surfactants that provide very low IFT, surfactants that act as hydrotopes to improve solubilization, and surfactant molecules for cleaning product uses at high or low temperature. In some cases, surfactants for formulations containing acids or alkali substances are needed.

2.1.2. Effect of Salinity

Scans of formulations are carried out using salts (e.g. potassium chloride, sodium chloride, calcium chloride, or calcium bromide) for the specific oil–water–surfactant systems used in microemulsion formulations. Studies of systems formulated with anionic surfactants exhibit a more significant salinity effect than systems formulated with nonionic surfactants. However, salinity also has an effect on nonionic surfactants. In the case of microemulsion systems with anionic surfactants, the addition of salts causes significant changes in the phase behavior of water–surfactant and oil–water–surfactant with addition of salts. The "optimal salinity" in the phase behavior of these systems is defined as the salinity at which a middle-phase microemulsion solubilizes equal amounts of oil and brine [25–27]. However, solubilization decreases for salinities higher than optimal salinity due to increased interfacial rigidity and curvature [28,29]. The existence of middle-phase microemulsions in equilibrium with excess oil and brine has been attributed to attractive interdroplet interaction and interfacial bending stress [28,30–32]. The highest efficiency of the microemulsion systems used to clean oily or dirty surfaces occurs at the salinity where the system reaches the minimum IFT and maximum solubilization.

2.1.3. Effect of Cosurfactant

In addition to the surfactants, substances such as alcohols can be used to act as cosurfactants to fine-tune or adjust the phase behavior of the brine–surfactant–oil systems to bring the microemulsion into the required experimental window of composition and temperature.

Addition of a short-chain alcohol (e.g. isopropanol and *n*-butanol) as a cosurfactant can increase the total interfacial area at low alcohol concentrations, thus increasing the solubilization [10,33] and decreasing the IFT. Figure 2.3(a) shows the dynamic IFT between the aqueous and oil phases used in a microemulsion formulation. One formulation is without alcohol and two formulations have a short-chain alcohol as cosurfactant. The results show that the IFT decreases with increasing length of the alcohol chain for this particular microemulsion system. The two formulations with cosurfactants show very low IFT from the initial contact with oil. The increase in short-chain alcohol concentration in the microemulsion system also reduce the IFT with time, as is shown in Fig. 2.3(b). Conversely, at high alcohol concentrations, phase separation occurs due to the increase in attractive interdroplet interaction. For this reason, only an optimal amount of alcohol is desired to formulate a microemulsion with maximum solubilization capacity. Increase of the alcohol chain length makes the system more lipophilic and reduces the attractive interdroplet interaction. The addition of an optimal amount of alcohol and salinity, together with the effect of the composition of the oil phase, may lead to the highest possible solubilization capacity of a given microemulsion.

2.1.4. Effect of Linkers

Lipophilic and/or hydrophilic linker additives could be used to increase the solubilization and modify the interfacial properties of the microemulsion [34–39]. These are amphiphilic molecules that segregate near the oil/water interface, near the hydrophobic surfactant tail (lipophilic linkers) or the surfactant head group (hydrophilic linkers) [34].

The lipophilic linkers are defined as molecules that, while present in the oil phase, orient along the surfactant tails and promote orientation of oil molecules further into the oil phase. The addition of a hydrophilic linker increases the space between surfactant molecules, a enabling flexible surfactant membrane, that leads to faster coalescence and solubilization kinetics [3]. Lipophilic linkers thus serve as a link between oil molecules and the surfactant tails [3,35,36]. Examples of lipophilic linkers are long-chain alcohols, such as alcohols with more than eight carbons. These molecules act as lipophilic linkers in microemulsion systems because they have an increased interaction with the oil molecules but do not adsorb at the interface. For alcohols having between four

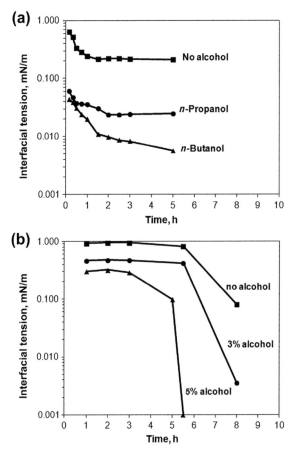

FIGURE 2.3 Effect of cosurfactants on the interfacial tension (a) between aqueous and oil phases of the microemulsion system and (b) between the hydrocarbon (petroleum) and three microemulsions formulated without alcohol, with 3 wt% alcohol, and 5 wt% alcohol.

and eight carbons, the alcohols behave as cosurfactants because they interact strongly with the oil but retain their adsorption at the oil–water interface. Small alcohols with less than four carbons show a cosolvent effect that helps decrease the surfactant–surfactant interaction.

The concept of hydrophilic linker was introduced later by adding a surfactant-like molecule that segregates near or at the oil/water interface, but due to its short tail offers little interaction with the oil phase [38].

Combinations of hydrophilic and lipophilic linkers can produce a surfactant-like system. The proper combination of lipophilic and hydrophilic linkers has been found to significantly increase the solubilization capacity for different oils [34,38,40]. The linker approach has been used to formulate microemulsions in applications such as environmental remediation and detergent formulations [41,42].

The linker molecules tend to extend the thickness of the interfacial zone over which the transition from polar to nonpolar phase takes place. The problem of the linker additives is that there is sometimes a need to add a large excess in order to have enough of it at the right place. This is a fractionation effect that results from the fact that very different molecules are likely to move where they find a better physicochemical environment [3,43].

Once the solubilization enhancement with linkers was understood, researchers started to work on synthesizing surfactants with spacer arms between the hydrophilic head and lipophilic chain to produce similar solubilization behavior compared with the combination of surfactants with linkers. These surfactants are called *extended surfactants*. In general, these molecules are longer than conventional surfactants and exhibit increased contact on both sides of the interface. However, if these molecules are too long, precipitation could occur. Examples of these molecules are alcohol ethoxylated propoxylated sulfates and alcohol ethoxylated propoxylated carboxylates [43–47].

2.1.5. Type of Oil or Solvent

The type and chemical composition of the oil and/or solvent used in microemulsion systems considerably influence their phase behavior and properties. When the oil molecules are solubilized in the aggregated core of the micelles, the micelles become swollen and their surface curvature tends to decrease, producing changes in the phase behavior [4,20]. The interaction between the oil molecules and the surfactant hydrophobic tail is a key parameter that must be considered in the selection of appropriate surfactants to obtain the desired microemulsion formulation.

In general, the oil and/or solvents are not alkanes composed of a single type of molecule. They are blends of molecules with a distribution of molecular weights and molecular configurations. Additionally, the oil and/or solvent can have a variety of chemical compositions that can be used in formulation and applications of microemulsions. In this case the EACN is used instead [15]. The EACN of an oil mixture can be deduced from its composition by a linear mixing rule based on molar fraction. The EACN of many oils has been measured [20]. The EACN concept is a very useful tool in the studies of microemulsion formulations. The EACN of an oil is a dimensionless number that reflects the "hydrophobicity" of oil. It is determined experimentally by comparing its phase behavior with that of a well-defined linear hydrocarbon in the same surfactant–oil–water system [5,14–16,48]. The EACN of a linear alkane is simply its carbon number. For example, the EACN of *n*-heptane and *n*-decane and *n*-dodecane are 7, 10 and 12, respectively. Oil molecules containing polar groups have low EACN because the presence of a polar group in the oil molecule reduces its EACN. Examples are ethyl oleate and C16–C18 triglyceride that have EACN of around 6 [5].

2.1.6. Temperature Effect

Temperature is an important variable that affects the performance of systems containing surfactants. The effect of changes in temperature on the phase behavior of surfactants in solution is very complex. The size of the micelles and the type of surfactant association change with temperature and affect the phase behavior of surfactant–water or surfactant–oil–water systems. Anionic surfactants typically become more hydrophilic as temperature increases, whereas nonionic surfactants present the opposite trend [49]. In the case of microemulsion systems formulated with only nonionic surfactants, the interaction between the hydrophilic part of the molecule and water increases as temperature decreases. For example, an increase in temperature or a decrease in ethylene oxide composition of the surfactant molecule has similar effects on formulations. In formulations with nonionic surfactants, their temperature effect is explained by the fact that the solubility of nonionic surfactants in water is due to hydrogen bonding, which is a very sensitive function of the temperature. Thus, increasing temperature may be expected to have a similar effect as increasing salinity [19,50,51].

To overcome the thermal sensitivity of the microemulsion formulations with either nonionic or anionic surfactants, a proper blend of anionic and nonionic surfactants could produce formulations with minimum sensitivity to temperature [49,51].

Another option to obtain thermally stable microemulsion formulations for applications at elevated temperatures, especially for temperatures higher than 150 °C, is to blend cationic surfactants with nonionic surfactants because some of these combinations exhibit excellent synergistic effects.

2.2. Properties of Microemulsions

2.2.1. Solubilization

Microemulsions solubilize immiscible water and oil systems. Solubilization in microemulsions results from the equilibrium coexistence of oil and water in the presence of surfactants and cosurfactants that form swollen micelles.

The increase of interactions between surfactant and oil and surfactant in water in a microemulsion system increases their solubilization. This can be quantified by obtaining the solubilization parameter (S_P). The solubilization parameter is the amount of oil solubilized in the core of the swollen micelles per unit mass of surfactant [3]. An alternative option to measure solubilization is to measure the minimum amount of surfactant necessary to produce a single phase in a surfactant–oil–water system. The curvature of the micelles decreases as the oil solubilization in the core of the micelles increases, up to a point where the HLD is zero or Winsor R is equal to one [3,20].

The solubilization capacity of microemulsion systems could be increased by using lipophilic linkers, hydrophilic linkers and/or extended surfactant molecules [3]. This approach is very attractive for many applications, including surface cleaning applications.

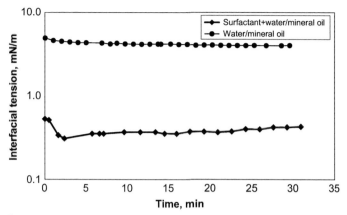

FIGURE 2.4 Interfacial tension of system used in microemulsion formulations for cleaning oily surfaces contaminated with oil-based drilling fluids.

2.2.2. Interfacial Tension

The reduction of IFT between the aqueous phase and the oil phase is important in the selection of additives to formulate the aqueous solution to maximize cleaning and water-wetting efficiency. The magnitude of IFT will be dictated by the addition of surfactant(s), cosurfactants and linkers over a range of compositions and temperatures corresponding to expected application conditions.

It is well established that very low IFT plays an important role in the cleaning efficiency and in oil removal from solid surfaces [52–54]. Selection of proper surfactants that reduce the IFT between the aqueous and oil phases plays a significant role in the overall effectiveness and ability of any fluid system to clean and perform in a prescribed environment [30].

Figure 2.4 shows an example of the dynamic IFT between mineral oil and the aqueous phase used in a microemulsion formulation for cleaning oil-based drilling fluids measured at 60 °C with a spinning drop tensiometer. The results presented in Fig. 2.4 show that when the surfactant–water system contacted the oil phase, the IFT dropped to 0.5 mN/m. Figure 2.4 also shows the IFT between the same oil and water as a point of comparison.

Another way to obtain the IFT is by using the theoretical relationship between solubilization ratios and IFT derived by Huh [55], which predicts that the IFT is inversely proportional to the square of the solubilization ratio at equilibrium. This approach enables the determination of the IFT without the need to measure IFT experimentally.

In a surfactant–water–oil system, IFT passes through a minimum and the solubilization parameter reaches a maximum at the optimum formulation condition. The minimum value of IFT and maximum value of S_P define the characteristics of an optimum system. These enable one to compare different systems from an overall point of view [5,21].

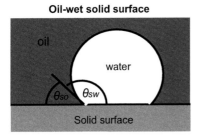

FIGURE 2.5 Schematic of contact angles of solid/water/oil interfaces.

2.2.3. Contact Angle and Wettability

The wettability of a surface is a characteristic that affects cleaning effectiveness, fluid displacement and solids mobilization. The contact angle of a water droplet on a surface, before and after exposure to different fluids, is an important consideration in evaluating a microemulsion cleaning fluid. When a liquid droplet is placed in contact with a flat solid surface, two distinct equilibrium regimes may occur: partial wetting with a finite contact angle (θ) and complete wetting ($\theta = 0$) [56]. Contact angle measurements are used to evaluate the ability of the fluids to change the surface wettability from oil-wet to water-wet and vice versa.

The determination of the equilibrium contact angle, described in Young's equation [57], requires very clean experimental conditions and depends on the liquid surface tension, the surface free energy of the solid and the interaction between the solid and liquid materials [56–58]. A deviation from the thermodynamic equilibrium condition results in contact angle hysteresis, which is observed by an advancing angle when the solid/liquid contact area increases and a receding angle when the contact area shrinks [56].

When microemulsion fluids contact solid surfaces, the interfacial free energy of the liquid–solid interface decreases, resulting in a reduction of the contact angle between the solid surface and the microemulsion. Figure 2.5(a) and (b) is the schematic description of this phenomenon showing the contact angles encountered in an oil-wet surface and a water-wet surface, respectively. If the oil-wet surface is exposed to a microemulsion cleaning fluid, the oily material is removed and the wettability changes from oil-wet to water-wet (the contact angle for the water is significantly reduced).

For practical applications, it is recommended that the contact angle measurement should be analyzed as a relative measurement compared to a baseline. One practical test procedure is to analyze the measured contact angle data as a measurement relative to a baseline of a clean, untreated glass slide. Figure 2.6(a) shows that the clean glass slide has a contact angle of 28° with the water. Then, to simulate contact of oily material or other nonaqueous fluid, the glass slide is exposed to oil containing oil-wetting surfactants for 10 minutes. The contact angle of the water, now 83° (Fig. 2.6(b)), is much higher than the native surface, thus proving that the previously water-wet surface has been transformed to an

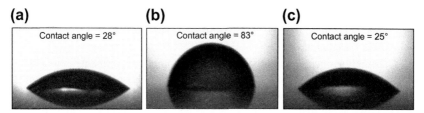

FIGURE 2.6 Contact angle between water drop and glass slide (a) clean glass slide, (b) exposed to the oil-based fluid and (c) exposed to sequence of oil-based fluid and microemulsion fluid.

oil-wet state after exposure to the oil-based fluid. Finally, the oil-wet surface is exposed to the microemulsion cleaner fluid for 10 minutes, resulting in a water contact angle of 25° as observed in Fig. 2.6(c). The surface has returned to its initial water-wet state. These results demonstrate that microemulsion formulation restores the wettability of the glass slide, as observed in this test.

3. BASIC PROCESS AND PRINCIPLES OF CLEANING SURFACES

The process of cleaning rigid surfaces composed of metallic or inorganic solids or "soft" materials, such as cloth or other flexible materials, begins with understanding the process of contamination of the surface, which involves the knowledge of the contaminating substance, the nature of the clean surface, and the contaminated surface.

The process of contamination, as discussed in this work, involves essentially the same principles as the process of wetting or coating of clean surfaces. Characterization of the physicochemical nature of the surface and of the contaminant(s) helps to understand the fundamental forces that hold or bind the contaminant to the surface [58].

Contaminated surfaces typically have forces of adhesion between the clean surface and the contaminant particles, and among the contaminant particles, which may appear to exist as a uniform fine layer or coating of the substance onto the clean surface. The forces that bond the clean surface and the contaminating substance may involve chemical adsorption and/or physical adsorption. The contamination process may have occurred over a long time scale (eons, as in the example of contamination of petroleum reservoir surfaces) or a relatively short time scale (years, months, weeks, days, or hours, as in contamination of wellbores by components of the fluid systems used in drilling operations). Examples of contamination of clean surfaces are found in the electronics industry, petroleum industry, environmental chemical and oil spills, food preparation, etc.

Many contaminants fall into the class of "oils" and/or organic substances that may exhibit amphiphilic, hydrophilic, hydrophobic, negatively, neutrally, or positively charged physicochemical properties. Interactions between contaminants and surfaces involve enthalpy contributions from intermolecular forces and entropy contributions related to steric ordering of particles or molecules at the surface or interface. In the process of contamination, these two contributions result in a lowering of the free energy of the system. The result is a surface with an adsorbed organic layer.

Thermodynamically, the process of adsorption of contaminants onto a clean surface involves a negative change in entropy (particles of the contaminant in the bulk fluid in contact with the surface are more randomly distributed and free to move about than the particles adsorbed onto the surface), and a negative change in enthalpy of the system, as shown in Eqns (2.8) and (2.9). The free-energy change for this process is also negative (Eqn (2.10)), indicating that the final state (oil-wet or contaminated surface) is favored over the initial state of the clean surface and contaminant particles randomly distributed in the fluid environment in contact with the clean surface.

$$\Delta S_{ads} < 0 \quad (2.8)$$

$$\Delta H_{ads} < 0 \quad (2.9)$$

$$\Delta G_{ads} = \Delta H_{ads} - T \Delta S_{ads} < 0 \quad (2.10)$$

One can view the process of cleaning of contaminated surfaces from a macroscopic perspective, in which we define macroscopic physical and/or chemical forces between the surface and the contaminating substance. One can also view the process from a microscopic or molecular perspective, in which we define microscopic physical and/or chemical forces between molecules on the surface or substrate and molecules (molecular assemblies or particles) of the contaminating substance.

The following discussion provides a visualization of the overall nature of the process of cleaning of contaminated surfaces, in general, and serves as a basis for thermodynamic analysis in terms of free energies and phase equilibrium of the system. The cleaning process necessarily involves thermodynamic equilibrium concepts and kinetic phenomena. The process of cleaning of a surface must occur within an acceptably short time frame, for practical and economic reasons.

Typically, forces of adhesion involve (1) electrostatic forces between the contaminant particles and the surface and/or (2) strong van der Waals forces between the contaminant particles and the surface and between contaminant particles/molecules in the layer(s) adsorbed onto the substrate.

To clean such surfaces one needs to overcome and neutralize the physical and/or chemical forces that bind the contaminant to the surface. There are essentially two types of methods: (1) use of mechanical energy (e.g. scraping, brushing, stirring, pressure washing, jetting, etc.) and (2) use of chemical solutions and/or solubilization agents (e.g. dissolving, solubilizing, etc.).

Mechanical options for removal of the contaminants may include shear (e.g. stirring) and/or introduction of fluid energy to dislodge the adhered particles (washing or jetting).

Chemical options include use of chemical substances that alter the balance of forces and cause the adhered particles to dislodge from the surface and become dissolved, dispersed and/or solubilized in a fluid phase.

Many cleaning processes involve combinations of mechanical energy and physicochemical methods, as in processes of washing clothes, automobiles, or other processes in which solvents or solutions of detergents are used in combination with mechanical agitation or pressure washing to effectively remove contaminants from the substrate material.

Each of those methods involves some amount and form of mechanical energy to improve the effectiveness of the cleaning process, to overcome surface-contaminant bonding forces, and/or to speed up the process.

The approach of using microemulsions as a cleaner has gained popularity, as their properties and formulation have been intensively studied in the last decades. The science of thermodynamically stable microemulsions has found a wide range of applications in cleaning processes, ranging from cleaning of soft materials to cleaning of rigid metallic surfaces and porous media in petroleum reservoirs. The special physicochemical nature of microemulsion systems, especially their quasi-zero IFT, enables these solutions to remove contaminants from surfaces essentially spontaneously, without the need for mechanical energy.

Thermodynamically, the process of cleaning a contaminated surface is essentially the reverse of the process by which the surface became contaminated. The free energy of the system after the cleaning process must be less than the free energy of the contaminated system. The entropy of the contaminant particles goes from the adsorbed state to a state in which they are incorporated into the microemulsion environment that is placed in contact with the contaminated surface. The entropy of the contaminant changes from a lower value (ordered state) to a higher value (randomly distributed within the microemulsion). In order for the free energy change to be negative for the process, Eqn (2.11) shows that the enthalpy change must be less than $T\Delta S$. Equation (2.12) defines the main contribution to enthalpy change as the energy required to produce a large increase in surface area of the contaminant, going from an adsorbed layer on the surface to a multitude of very small, quasi-molecular scale droplets or assemblies incorporated into the microemulsion environment.

Thermodynamics of this spontaneous cleaning process is such that the free energy of the system is negative, i.e.

$$\Delta G^* = \Delta H^* - T\Delta S^* < 0 \qquad (2.11)$$

which implies

$$\Delta H^* < T\Delta S^* \qquad (2.12)$$

And, because the enthalpy change is mainly from energy to increase surface area,

$$\Delta H^* = \gamma \Delta A^* \qquad (2.13)$$

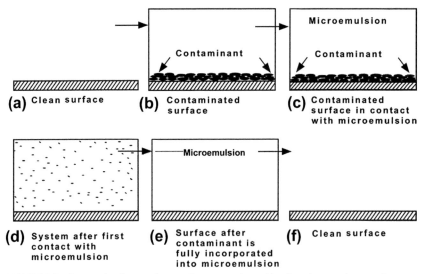

FIGURE 2.7 Progression from a clean surface to a surface with adhered contaminant to the same surface upon contact with a microemulsion.

where
ΔA^* = increase in surface area of the adsorbed contaminant, going from "flat sheet" adsorbed state to the state of many small particles or assemblies of contaminant molecules incorporated into the microemulsion.

γ = IFT between the contaminant substance and the bulk microemulsion fluid in contact with the substrate.

The surface area of the contaminant cannot be negative. The surface area actually increases greatly as the "flat sheet" of adsorbed contaminant breaks down into many small molecular-scale assemblies incorporated in the microemulsion. Therefore, the minimum free energy of the system corresponds to the condition where the IFT approaches zero, i.e. $\gamma \approx 0$.

To visualize the process, Fig. 2.7 shows a progression from a clean surface to a surface that has adhered particles contaminating it, and then to the same surface upon contact with a microemulsion, with following steps in which the contaminant is spontaneously incorporated into the microemulsion, with no addition of mechanical energy. For our purposes, we will consider the contaminant particles to be of an oily nature that adhere to the surface via van der Waals forces.

The contaminant material is spontaneously drawn into and incorporated with the microemulsion phase with no addition of shear or mechanical energy. This makes "spontaneous" microemulsion cleaning a very useful method for cleaning of surfaces in which it would be difficult or impossible to provide mechanical energy, e.g. in cleaning of porous media in petroleum reservoirs and wellbores.

4. SURFACE CLEANING AND CONTAMINANT REMOVAL WITH MICROEMULSIONS

Cleaning of surfaces with microemulsions is a complex process that depends on several factors, including the concentration and composition of the wash solution, temperature, mechanical energy, and the nature of the substrate to be cleaned [59].

The substrates could be either (1) hydrophobic surfaces (oil-wet) such as polypropylene, polyethylene and polystyrene that have contact angle to water in the range of 90–100° or (2) hydrophilic (water-wet) surfaces such as glass surfaces that have contact angles to water in the range of 20–30°. Surfaces such as heterogeneous rock formations present mixed wettability. The surface contaminants could be liquids or solid particles and can have different degrees of adhesion to the surface, depending on their composition and properties.

Key aspects that determine the reduction of the adhesion between the soil and the substrate are the particle and/or liquid–substrate interactions, cleaner–contaminant interactions, and cleaner–substrate interactions. Mechanical energy and flow dynamics are additional aspects that play an important role in the cleaning process.

The primary mechanism involved in the cleaning process with microemulsions is solubilization of oily materials into the core of the micelles to form swollen micelles. Another mechanism that is also observed is soil roll-up and microemulsification of the oily material [52]. The IFT between the oily material and the washing or cleaning fluid plays an important role in the cleaning process. Microemulsion systems that exhibit ultralow IFT in the Winsor III phase region, passing through a minimum at the washing condition of the surfactant–soil–water system, may be excellent cleaners.

5. DESIGN OF MICROEMULSION CLEANERS AND EVALUATION TECHNIQUES

The design criteria of microemulsion fluids depend on the specific application. Some of the requirements that determine the design of a microemulsion include the cleaning time, solubilization or removal of the oil from the surface, and resulting wettability of the surface. Depending on these requirements, the design could be an optimum formulation or a Winsor III, Winsor IV or a Winsor I on the borderline of Winsor III. This last system could be designed to form an in-situ microemulsion during the cleaning procedure.

Formulation of microemulsions with these desired characteristics requires proper choice of a surfactant blend plus other additives, such as cosurfactant(s) and linkers to form the appropriate fluid mixture that is most efficient for a broad range of conditions (types of oily material to be cleaned, salinities, O/W ratio, temperatures, etc.). The formulation studies that can be used to select the best microemulsion for cleaning applications include a solvent or an oil with EACN similar to the material to be cleaned, together with freshwater or brines (NaCl,

CaCl$_2$, NaBr, and CaBr$_2$), and surfactant blends over a temperature range that covers the application conditions. Short-chain alcohols and linkers are optional additives in the formulation.

The physicochemical characterization of the microemulsion systems includes studies of phase stability, phase behavior, and molecular interaction [60]. The mechanism of micellar solubilization and microemulsion characterization can be studied by methods such as nuclear magnetic resonance (NMR), ultracentrifugation, X-ray diffraction, small-angle neutron scattering, and differential scanning calorimetry [60–63]. NMR is a very useful technique in the studies of phase behavior and microstructure. Formulation scan is a practical way to study the phase behavior of the surfactant–water–oil microemulsion systems. In addition to the formulation scans, the IFT, wettability and laboratory procedures adapted for the specific applications are used to evaluate the effectiveness of these microemulsion systems.

IFT between liquids can be measured by a number of methods [64]. Dynamic IFT measurements have been recognized to be important for understanding interfacial processes in industrial operations that involve multiphase systems [65], such as microemulsion cleaning processes. In industrial operations that involve liquid–fluid interfaces, the composition of the fluid at the interface is constantly refreshed and does not reach equilibrium.

Techniques for measurement of the dynamic IFT include the maximum bubble pressure, growing drop (bubble), oscillating jet, and pulsating bubble methods. If the IFT of the measured system reaches ultralow values such as those encountered in microemulsions used for drilling fluid removal (lower than 10^{-2} mN/m) from rock and metal surfaces, the most appropriate techniques are spinning drop tensiometry and surface light scattering [64–66].

The maximum bubble pressure technique is useful for measuring IFT between fluids that have similar density. Under these conditions commonly used techniques for measuring IFT fail, because they depend on the existence of a significant density difference. An example where this technique has been used is the measurement of IFT between bitumen and water which have an extremely small (<0.001 g/cm^3) density difference [67].

6. MICROEMULSION CLEANING APPLICATIONS

The distinctive properties of microemulsions (high solubilization of oil, low IFTs, and spontaneous formation) make these fluids very attractive for a variety of cleaning applications. Microemulsions have been used in industrial cleaning processes and household cleaning applications. Some of the applications include domestic and industrial laundry, cleanup of wastewater, cleaning of contaminated soil and various applications in the oil and gas industry.

In each of the following examples of microemulsion cleaning applications, the hydrocarbon or oily materials are solubilized in the aggregated cores of the micelles, producing swollen micelles and, as a consequence, its surface curvature tends to decrease, producing changes in the phase behavior, as shown

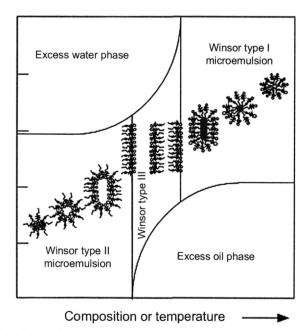

FIGURE 2.8 Effect of changes in composition or temperature on microemulsion phase behavior.

in Fig. 2.8. Depending on the application conditions, phase changes can be triggered by temperature or by composition changes. In some applications, the treatment solution is designed to be a microemulsion system with the phase behavior bordering between Winsor I and Winsor III, such that when the solution contacts the oily contaminant, the oil is solubilized into the micelles by a diffusion mechanism and the system transitions to the Winsor III-type microemulsion. In other applications, the treatment solution may be designed as an oil-based Winsor II-type microemulsion system on the borderline of the Winsor III system. When the solutions are in the Winsor III condition (or optimum formulation), the IFT drops very dramatically to near zero, which facilitates incorporation of oil or water into the amphiphilic structures.

The use of microemulsion fluids is an emerging technology for cleaning synthetic and hydrocarbon material in the oil and gas industry. Applications range from surface to downhole applications and from static conditions to dynamic displacement in the turbulent flow. Some of the microemulsion cleaning applications include the following:

- Removing oily material from oil-contaminated drilling cuttings brought to the surface during the drilling operation of the well.
- Cleaning metal surfaces of the pipes and rock surfaces of the wellbore during the displacement of brine-in-oil emulsion drilling fluids (typically called oil-based mud (OBM) or synthetic-based mud) to water-based fluids during the wellbore construction of the well.

- Cleaning wellbores and removing the filter cake formed by the brine-in-oil emulsion drilling fluids at the surface wall of the wellbore after the drilling process to assure the well produces the expected crude oil production.
- Removing hydrocarbon deposits, sludge or viscous crude oil emulsion, and formation damage due to drilling fluid invasion in the near-wellbore region of the production zone of oil and gas wells.

The following discussion focuses on the detailed mechanisms and examples of the aforementioned applications.

6.1. Cleaning of Oil-Contaminated Drill Cuttings

Drilling fluids are specialized fluids that are circulated down the well during the rotary drilling operation to perform various functions, such as cooling and lubricating the drilling bit and flushing the rock cuttings (drill cuttings) to the surface. Offshore and onshore disposal of drill cuttings generated from wellbores while drilling with oil-based or synthetic-based muds is still an expensive and challenging process. In some areas, such as the North Sea, legislation established a maximum allowable discharge of <1% oil. This regulatory action is a consequence of the environmental impact of oil-wet drill cutting accumulations on the ocean floor ecosystem, even with synthetic-based drilling fluids that pass all required environmental tests.

Drill cuttings are typically agglomerates of rock fragments, crude oil and drilling mud. In general, oil is either coated onto the drill cutting surfaces (surface oil) or trapped/occluded inside the porous drill cuttings clusters due to high capillary forces, as shown in Fig. 2.9. To decontaminate these drill cuttings, surface oil and trapped oil must be thoroughly removed.

Various mechanical and chemical processes, including microemulsions, could be applied to convert these oil-contaminated drill cuttings to an acceptable clean form. Microemulsions could treat oil-contaminated drilling cuttings simply by mixing the solids to be treated with an aqueous wash solution that

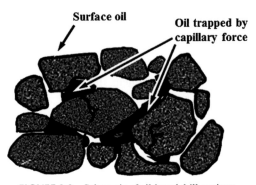

FIGURE 2.9 Schematic of oil-based drill cuttings.

TABLE 2.1 Interfacial Tension between Aqueous Phase and Oils System for Cleaning of Drill Cuttings

	Interfacial Tension, mN/m	
Type of oil	With 2 wt% surfactant	With 0.5 wt% surfactant
Diesel	1.4×10^{-4}	2.0×10^{-3}
Mineral oil	4.6×10^{-4}	3.3×10^{-2}
Paraffin oil	3.6×10^{-4}	3.1×10^{-3}

is a highly diluted microemulsion. This type of wash could offer high cleaning efficiency while requiring minimum energy and mechanical input. The oil is liberated from the drill cuttings into the wash solution by the reduction of the oil/wash solution IFT.

Microemulsion formulations that include surfactants, freshwater or seawater, base-oils used in drilling fluids, and water softener are studied with the object of formulating wash solutions for treating oil-contaminated drill cuttings. For this particular application, surfactant blends formulated with anionic–nonionic surfactant mixtures and with commonly known water softeners, such as sodium silicate, were used.

Table 2.1 presents the IFT measured between various oils and a wash solution selected from studies that included a number of surfactant mixtures formulated with seawater. The tests were performed with common base-oils used in drilling fluids. According to Table 2.1, the surfactant mixture used produced low IFT in a range of 10^{-3} and 10^{-4} mN/m with surfactant concentration between 0.5 and 2 wt% in the wash solution. Studies of the phase behavior performed with this oil–water–surfactant system formed a Winsor III microemulsion.

The wash solution used in the tests shown in Table 2.1 was used to treat various oil-contaminated cuttings with oil content ranging between 10 and 13 wt%. The results of the wash process showed a reduction of oil retention on the cuttings (ROC) to the order of 3 wt%. The same formulation with the addition of a cosolvent reduces the ROC to around 1 wt% (Fig. 2.10).

6.2. Wellbore Cleanup During Displacement of Oil-Based Drilling Fluids to Water-Based Fluid

When displacing and completing oil and gas wells, contaminants, such as oil-based drilling fluids, drilled solids, and other debris found in casing and risers, should be removed quickly and efficiently. Fluids incompatibility is a major issue when oil and gas operators have directly displaced the oil-based drilling fluid from a wellbore with an aqueous fluid such as brine and/or cement after running a casing string.

FIGURE 2.10 Drill cuttings (a) before treatment and (b) after washing with microemulsion fluid.

The vast chemical dissimilarity between the oil- and water-based fluid systems could produce a highly viscous fluid–fluid mixture when oil-based drilling fluids and water-based fluids come into contact with each other. This can result in spikes in viscosity, fluids channeling, poor cleaning, settling of solids, insufficient water-wetting, etc., decreasing the displacement efficiency. These inefficient displacements could lead to poor cement bonding, squeeze jobs, communication between zones and, in some cases, blowouts [68–71].

Microemulsion technology has been used as a spacer fluid between the oil-based drilling fluids and brine completion or between the oil-based drilling fluids and cement slurry to resolve this problem. This type of spacer prevents the aforementioned problem and cleans the oily debris from the casing and water-wets the metal surfaces and rock formation to promote good cement bonding [72,73].

Selecting the proper composition of the microemulsion spacer that meets the requirements of oil-based drilling fluid displacement efficiency and changes the wettability of the surface casing from oil-wet to water-wet is typically dependent on the drilling fluid type, composition, and wellbore conditions [73,74]. Efficient displacement is defined as the complete removal of the oil-based drilling fluid from the wellbore and from the metal surfaces, as illustrated in Fig. 2.11. Figure 2.11 shows a metal sleeve surface exposed to the oil-based drilling fluid for 15 min at 100 rpm and at 65 °C, and the same sleeve after exposure for a few minutes to an efficient microemulsion spacer. The metal sleeve coated with drilling fluid formulated with synthetic oil is shown below in Fig. 2.11(a). Visual inspection of the sleeve in Fig. 2.11(b) provides an indication of the cleaning effectiveness of microemulsion spacers. The results indicate >99.5% of the drilling fluids were removed and the metal surface is rendered water-wet.

Similar cleaning efficiency was obtained when this microemulsion formulation was used to displace oil-based fluid and to clean the casing, riser and drill pipe during the wellbore construction of oil wells. Figure 2.12 shows a drill pipe covered in drilling fluids before displacement in an oil well. This closely resembles

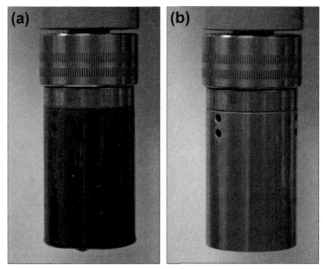

FIGURE 2.11 The stainless steel sleeve after (a) exposure to synthetic-based mud (SBM) and (b) treatment with the microemulsion spacer.

FIGURE 2.12 Photographs of a drill pipe (a) before and (b) after displacement with the microemulsion spacers.

the situation of the metal surface prepared in Fig. 2.11(a). Figure 2.12 also shows the drill pipe after a single displacement using the microemulsion formulation. Observation reveals results very similar to those obtained in the laboratory.

To avoid channeling problems and to remove a majority of the debris from the rock and metal surfaces during the fluids displacement, the spacer that

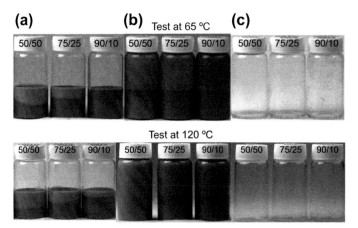

FIGURE 2.13 Cleaning test at various ratios of spacer/OBM in glass vials: (a) initial state, (b) after 16 h aging, and (c) after slight rinse.

contacts the oil-based fluids must have rheological properties, such as viscosity, relatively higher than those found in the drilling fluids. This is achieved by adding a viscosifier substance to the spacer microemulsion fluid. To assure total removal of the oil and to completely water-wet the surfaces, the viscosified spacer is followed by a second microemulsion spacer formulated in brine that completes the cleaning and renders the surfaces water-wet.

A number of tests are used to evaluate the cleaning efficiency of the microemulsion systems, including rheology of spacer/drilling fluid blends, sleeve cleaning tests and vial cleaning tests. Figure 2.13 shows an example of vial tests where a viscosified microemulsion spacer is used to remove oil-based drilling fluid from glass surfaces. A series of vial tests that simulate the drilling fluid/spacer interface are carefully prepared with oil-based drilling fluid and the spacer fluid at ratios between 50/50 and 90/10. Figure 2.13 shows the appearance of the vials after completing the test procedure that involves mixing, aging for 16 hours at selected temperature, pouring out and a slight rinsing with water. The appearance of the vials reveals the cleaning and water-wetting effectiveness of the microemulsion spacer fluid for cleaning of surfaces wetted with oil-based drilling fluid. The results of the tests at 65 and 120 °C were very similar, because the formulation studies defined a surfactant blend–water–oil system that maintains useful properties in a broad range of temperature.

6.3. Near-Wellbore Cleaning in Oil and Gas Wells

Microemulsion fluids have been successfully used to effectively resolve the persistent problem of near-wellbore formation damage. The term "formation damage" in the oil and gas industry refers to any process that causes a reduction in the natural productivity of hydrocarbons in the porous media in an oil and/or gas well. This is a by-product of the drilling, completion, and production process and can be attributed to many factors. In open-hole (OH) and cased-hole (CH) wells, hydrocarbon flow may be impeded by various formation damage mechanisms caused during the drilling and completion process and/or the production phase of the well. Formation damage may originate from fluid invasion into the surrounding rock during the drilling operations, organic deposition from the reservoir hydrocarbon system, oily debris left downhole, adsorption of additives such as surfactants and polymers used as operation fluids, and/or fluids incompatibility. The damage caused from incompatibility between in-situ reservoir fluids and any of the operation fluids typically results in in-situ water-in-crude oil emulsions or sludge formation. These fluids include drilling fluids, completion fluids and stimulation fluids.

The high oil solubilization, high diffusion through porous media, and the reduction of IFT between organic and aqueous phases to near zero make microemulsions excellent candidates for removing formation damage. These systems are excellent choices for cleanup and removal of synthetic or oil-based drilling fluids filter cake in OH completion applications. Formulations have also been developed for CH perforation applications, as well as for postdrilling remediation treatments to remove formation damage around the perforation or fracture zones.

The selection of the microemulsion fluid technology for near-wellbore cleaning depends on the specific application. For example, in a CH completion well, for either prevention or remediation of formation damage, a Winsor IV-type microemulsion performs best. In OH wells, where a uniform filter cake cleanup is required, the best solution is to pump downhole a Winsor I system approaching the transition to optimum formulation that will form an in-situ microemulsion when it contacts the filter cake. This is considered to be the best method to achieve effective OBM filter cake removal throughout the hole section.

6.3.1. Removal of Oil-Based Fluid Filter Cake in OH Completion Wells

Oil-based drilling fluids are emulsions formulated with various additives that include emulsifiers, viscosifiers, lubricants, and solids (e.g. calcium carbonate and barium sulfate) to adjust the fluid density. During circulation of the drilling fluids under pressure from surface to downhole, leak-off or a small volume of filtrate of drilling fluids is expected to pass through the porous medium, leaving a filter cake around the wellbore wall of the porous medium. The filter cake includes a combination of the drilled solids, the solids that have been added to the drilling fluid, chemical additives, brine and/or oil. The formation of a filter cake (typically a few millimeters in thickness) is extremely important for controlling fluid losses to the rock formation during the drilling process; however, the filter cake must be removed after the drilling process is finished. Failure to remove the filter cake impedes the expected hydrocarbon production. An additional near-wellbore damage (reduction of hydrocarbon production) could result when poor filter cake formation allows filtrate invasion deep into the rock formation.

The cleaning time required to affect good wellbore cleanup depends on the filter cake properties, such as cake thickness, toughness, slickness and permeability, as well as the downhole temperature. In the case of oil-based filter cake, the cleaning fluid is pumped and allowed to soak for a minimum of 24 or 48 hours, and sometimes longer, because it is difficult to predict the exact properties of the filter cake.

One-step oil-based filter cake cleanup technology uses a single-phase microemulsion and conventional acids, in a single blend, to solubilize the oil into the microemulsion, reverse the wettability of the filter-cake solids, and simultaneously decompose its acid-soluble components [75,76]. Reversing the wettability of the filter cake, using surface-active chemistry, facilitates acidizing by preventing a sludge that could form between the acid and the emulsified cake and by making acid-soluble particles unavailable to unspent acid. Besides the advantage of reduced skin damage in the well, increased hydrocarbon recovery and/or increased water injection rates, a "one-step" near-wellbore cleanup method could save an operator oil company valuable rig time.

To select an efficient formulation for filter-cake cleanup, phase behavior studies that include a solvent or base oil used in oil-based drilling fluids, together with brines (NaCl, $CaCl_2$, NaBr, and $CaBr_2$), and surfactant blends were evaluated over a temperature range of 25–90 °C. Short-chain alcohols and acids are optional additives in the formulation and were evaluated in the phase diagrams. Phase behavior was studied using an oil/aqueous phase ratio of 1–4. The oil phase used in the studies includes solvents and base-oils used for oil-based drilling fluids. The data obtained from the phase behavior studies were used to build phase diagrams, allowing better understanding of capabilities and possible performance of the fluid formulated with a particular surfactant–oil–brine system.

Figure 2.14(a) shows an example of phase behavior of a system used for filter-cake cleanup. In this particular case, the oil–water–surfactant phase behavior study used the same type oil found in the oil-based drilling fluids, to assure good oil solubilization when the cleanup fluid contacts the filter cake. The example shows a system that is very sensitive to changes in salinity, such as the systems formulated with only anionic surfactants. The phase behavior shows that the increase in salt concentration in the fluid formulations causes the fluids to approach close to the boundary of WI–WIII. Based on this observation, two formulations from the same system would perform very differently, insofar as cleaning the same material, if they are formulated at very different salinities. Figure 2.14 shows photographs of the discs with an oil-based filter cake before (Fig. 2.14(b)) and after the cleaning tests with Formulation 1 (Fig. 2.14(c)) and Formulation 2 (Fig. 2.14(d)). Formulation 2, formulated in 15% $CaCl_2$, has

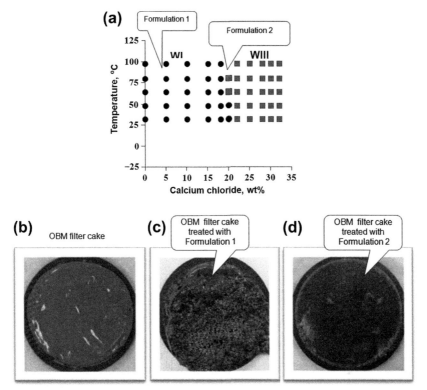

FIGURE 2.14 Cleaning performance of two fluids selected from the surfactant–oil–brine system presented in Fig. 2.14(a).

much better cleaning efficiency than the Formulation 1 with 3% $CaCl_2$, which is located much farther away from the transition from Winsor I to Winsor III-type microemulsion than is Formulation 2. The clean state of the disc indicates that the oil of the filter cake was solubilized, turning the solids water-wet. Once the calcium carbonate solids used in the fluids became water-wet, they were dissolved by the acid (around 10% by weight in the fluid).

Regain permeability tests are performed to evaluate the ability of the microemulsion treatment fluid to remove oil-based filter cake and to remove damage caused by wettability changes and in-situ water-in-crude oil emulsion formation. Figure 2.15(a) shows a schematic of a permeability test setup in a permeameter equipment. The tests are run using ceramic discs as porous media. The test procedure begins with measurement of the initial permeability (K_i) to the brine. The following step is the formation of a filter cake on a ceramic disc, followed by treatment with the microemulsion soak solution for a specified period of time at a specific temperature. After the soak period, the final permeability (K_f) was measured at various flow rates. Then, the cell is opened and the residual filter-cake solids are assessed for water-wetness and dispersion ability.

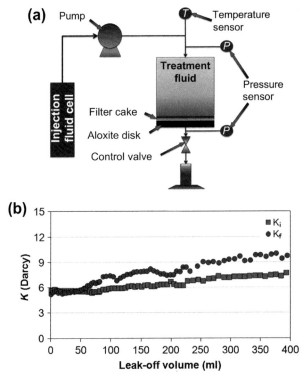

FIGURE 2.15 Regain permeability test (a) schematic of permeability tests setup, and (b) regain permeability results.

Figure 2.15(b) shows the permeability data recorded during the tests using the formulation selected from a point in the phase diagram (Fig. 2.14(a)) that corresponds to a borderline transition from Winsor I to Winsor III. The results show that the final permeability, measured after the soaking, was even higher than the initial permeability, which is a desirable result. The formulation removed the oil and acid-soluble solids of the nonaqueous fluid filter cake, turned the residual solids from oil-wet to water-wet condition, and cleaned the porous media (ceramic disc). These results are an indication of hydrocarbon production enhancement that could be obtained by using microemulsion cleaning fluids in oil and gas wells.

6.3.2. Removal of Formation Damage in CH Completion Wells

Other downhole cleaning application of microemulsion fluids is in CH wells where the crude oil production may be impeded by formation damage. Some of the causes of formation damage or reduction of crude oil production encountered in wells completed with CH include wettability changes, in-situ emulsion formation or sludge, and water blockage [77,78].

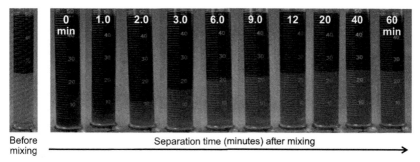

FIGURE 2.16 Fluids separation in emulsion tests using crude oil and a Winsor IV system fluid.

An approach to treat these wells is to use a microemulsion fluid that removes formation damage in the near-wellbore region [78]. The treatment fluids diffuse into the perforated rock matrix and spontaneously solubilize oil and remove the blocking material in the pore rock. The fluid simultaneously water-wets and fluidizes solids in the damaged porous media to prepare them for easy removal during production operations. For this particular application the Winsor IV-type systems are used to have a fast cleaning action. In addition to the ultralow IFT, low contact angle, and good cleaning requirements, the system used for this particular application needs to provide a fast separation when evaluated in emulsion tests with crude oil. This type of test is performed by evaluating the treatment fluid in emulsion risk tests.

The risk of water-in-oil emulsion formation in porous media is evaluated using classical bottle tests. These tests are performed using a 50/50 blend of crude oil/microemulsion fluid at the required temperature for applications. The mixtures are vigorously mixed by hand-shaking in a graduated cylinder. The cylinder is then allowed to rest at a predetermined temperature while separation of the fluids is observed.

Figure 2.16 shows the results of an emulsion test using crude oil and the microemulsion fluid treatment. A total separation of the fluids was achieved in about 20 min. This result indicates that this particular microemulsion evaluated is effective in preventing and removing formation damage caused by blockage sludge formation in the porous media.

6.4. Other Cleaning Applications

Other cleaning applications of microemulsion fluids that have been reported include cleaning of wastewater, fabrics and textiles, soil contaminated with jet fuel, frescoes and paintings that have been polluted, fracture paths in shale gas formation, and sweeping of oil in petroleum reservoirs.

In this section, the term "oily" is used to mean not only nonaqueous liquids, with hydrocarbon character, such as mineral oils, manufactured solvents, crude oil or petroleum, refined petroleum, and vegetable oils, but also semisolid materials such as tar, asphalt, grease deposits, etc.

6.4.1. Wastewater Cleaning and Microemulsion Froth Flotation

Microemulsion technology has been shown to be effective for cleanup of oily wastewater [79,80] in batch and continuous processes. Experimental results indicate maximum oil removal from water to occur at surfactant concentrations at or near the critical composition at which microemulsions form in the system. The most favorable condition corresponds to Winsor III microemulsion. Water purification by microemulsion froth flotation has been studied for removal of contaminants such as motor oil, diesel, chlorohydrocarbons, carbon black, fine mineral particles, etc. The process typically involves identification of an appropriate surfactant or surfactant blend for the "oil"/water/surfactant system, followed by the studies of microemulsion formation in response to salinity of the aqueous phase, and determination of the optimum air bubbling mechanism to cause the contaminant "oil" to float to the air–liquid interface at which a surface skimming process removes the froth and contaminant. This froth flotation separation process involves various phenomena, and variables, such as attachment of the oily droplets or swollen micelles of microemulsion to air bubbles, the size distribution of the oily droplets and air bubbles, dynamic IFTs, froth stability, and ability of the froth to suspend the floated oily droplets [79]. Experimental results favor use of small quantities of surfactant in the system near the critical composition for microemulsion formation. This method can be useful for purification of wastewater and recovery of valuable trace contaminants.

An interesting application is the use of microemulsions in the froth flotation process to cleanup wastewater [79,80]. In this process the surfactant interacts with the air bubbles and with the dispersed oily droplets or dissolved oily material in the wastewater. Thus, swollen surfactant micelles form in the contaminated aqueous phase. Figure 2.17 shows that the air is introduced into the system via a bubbler device. The surfactant concentrates at the air/water interface with hydrophobic part(s) oriented toward the inside of the air bubble and the hydrophilic part orients in the aqueous phase. With the oily droplets, the hydrophobic part of the surfactant orients toward the center of the oily drop and the hydrophilic part orients toward the aqueous phase. The dispersed oily droplets are attracted to the rising air bubbles and accumulate in a foam or froth layer at the water surface. Skimming of the surface removes the froth.

This method can also be applied to cleanup of solid particles such as sand or soils, by contacting the contaminated sand or soil with an aqueous solution containing surfactant that will form a Winsor III microemulsion. The wash liquid would then be separated from the solids and subjected to a froth flotation step in which the contaminant would be removed by air injection into the aqueous surfactant system containing the contaminant "oil".

6.4.2. Microemulsion Cleaning of Contaminated Soil and Groundwater

The contamination of soils and groundwater with volatile and/or nonvolatile organics from underground storage tanks, spills, and improper waste disposal

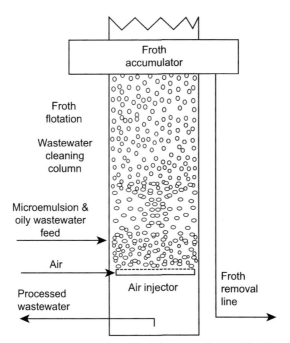

FIGURE 2.17 Wastewater cleaning with surfactant or microemulsion froth flotation.

presents a major remediation problem [81]. Various publications have discussed the application of microemulsions and mechanisms of contaminant removal at various conditions (temperature, salinity, and surfactant concentration). One particular example is the removal of chlorinated hydrocarbons used in cleaners and degreasers products that are commonly found in the groundwater [82].

Another variant of the oily soil cleaning process may involve the injection of a microemulsion solution at an injection well into a bed of contaminated soil and withdrawal of the injected solution plus oily contaminant via a removal well, such that the solution migrates from the injection site to the removal site, flushing and cleaning the contaminants from the channels in the porous media. Figure 2.18 shows a schematic of the underground contaminated soil treatment and displacement out of the ground.

An example is the use of microemulsion technology to clean soil contaminated with tetrachloroethylene. A field demonstration showed that this type of fluid is able to considerably reduce the percentage of tetrachloroethylene in soil. After 10 pore volumes of flushing, the overall contaminant in the underground soil was considerably reduced when compared with results from using conventional technologies [83].

Recovered contaminant and aqueous solution of surfactant could be fed to a process unit in which air injection can be used to create a froth to concentrate the microemulsion and the oily contaminant at the top surface of the liquid.

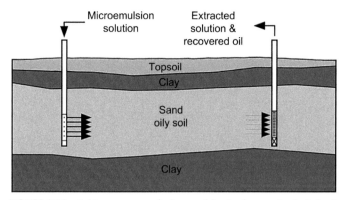

FIGURE 2.18 Subterranean use of microemulsion to clean contaminated soil.

Examples in which this type of process has been used include the decontamination of soils near fuel-spill sites or contaminated soils from fuel storage tank leakage.

6.4.3. Microemulsion Cleaning of Textiles

An important application in which microemulsions are useful involves cleaning of oily contaminants from fabrics and textiles. The contaminants may be either particulate soils or liquids. Both types of contaminants adhere to the substrate textile as a consequence of van der Waals and related interactions between the contaminants and substrate [84].

Studies have shown that, for certain cases, traditional cleaning of fabrics and textiles using volatile organic solvents or surfactant solutions can be replaced with specially-formulated environmentally friendly microemulsions in aqueous solutions. Microemulsion cleaning of fabrics and textiles is linked with the general notion of detergency.

The primary cleaning mechanism of microemulsion is the solubilization–emulsification mechanism at the soil–detergent solution interface [52]. This mechanism is directly influenced by the phase behavior of the corresponding oil–water–surfactant system used. The microemulsification is influenced by the variables encountered in the process, such as the composition and surface properties of the fabric, amount and type of soil or contaminants, water hardness and temperature [52,85].

Cleaning of fabrics involves use of appropriate surfactants to create microemulsions at the optimum formulation conditions or near the transition to a Winsor III-type microemulsion, during the washing or cleaning process. This very low IFT condition facilitates solubilization and incorporates the oily contaminant and subsequent removal of the contaminant from the fabric by flushing the fabric with an aqueous solution. In application, the microemulsion cleaning solution is reusable multiple times, because the performance and phase behavior of the cleaning solution change very little in each use, due to the relatively large amount of presolubilized oil and the small amount of oily contaminant removed in each cleaning.

In some cases, stains on fabrics and textiles can be removed by application of an appropriate pretreatment using a concentrated surfactant system that forms a microemulsion with the oily contaminant on the fabric. The concentrated surfactant needs to remain in contact with the oily contaminant for a sufficient time to diffuse into and mix with the oily contaminant, prior to flushing of the stained area with a clean aqueous solution.

6.4.4. Microemulsion Cleaning Using Nonaqueous Solvents

Another interesting application of microemulsions for cleaning is their use in systems that do not contain water. Some applications [86,87] use substances like carbon dioxide (CO_2) as the continuous phase, instead of water, to effect a dry-cleaning process for situations where materials being cleaned would be damaged by contact with water. Special surfactants are used for these systems and cleaning applications. Surfactants for CO_2 microemulsions have been formulated from special copolymers composed of polystyrene segments (CO_2-phobic) and modified fluorine-substituted acrylonitrile segments (CO_2-philic). Other surfactants that have been studied for CO_2 textile cleaning applications include di-n-octylamine [88,89]. One application is in the dry-cleaning of special fabrics. Other applications are the cleaning of electronic components and assemblies that would be damaged by contact with water.

6.4.5. Microemulsion Cleaning of Building Exteriors

Microemulsions can be used for cleaning of the exterior surfaces of buildings. Tourists and travelers may recall the sooty historical buildings, cathedrals and castles in parts of northern cities in America, Europe and the Old World caused partly by decades or centuries of exposure to smoke and soot from chimneys and heaters that used coal or other fuels that emitted particulates that resulted in the dingy gray look of originally beautiful stone and masonry cathedrals, palaces and houses of government. Prior to the development of microemulsion technology, such buildings and monuments were subjected to cleaning procedures that used harsh sandblasting and/or pressure washing, causing more harm than good by permanently removing layers of the building surfaces. With microemulsion technology, such surfaces can be cleaned by using effective, fast-acting microemulsion solutions that penetrate the pores, cracks and crannies of the stone edifices to spontaneously form microemulsion nanostructures that solubilize and incorporate the contaminants, and facilitate their subsequent removal by gentle washing of the surfaces with clean water [90].

6.4.6. Microemulsion Cleaning of Frescoes and Artwork

Another application involves use of microemulsions to clean frescoes and paintings that have become coated with pollutants or damaged by vandals with graffiti. Special microemulsion systems have been able to remove the contaminants with minimal damage to the original artwork [91].

Studies in microemulsion cleaning proved that highly insoluble acrylic polymers and organic materials employed in works of art and architecture can be solubilized in microemulsions. An example of this application is the complete cleaning of an acrylic-contaminated painting successfully achieved by using a microemulsion system. The microemulsion system reduced about 95% of the total amount of the organic phase (1% w/w) on the contaminated piece of art. The microemulsion was particularly effective for the cleaning of a secco painting, a fragile painting difficult to clean using conventional methods, as in the case of the restoration of the wall painting by Vecchietta in the Sacristy of Santa Maria della Scala, in Siena, Italy [91].

6.4.7. Microemulsion Cleaning of Crude Oil Reservoirs

A very interesting challenge for microemulsion cleaning is the field of enhanced recovery of hydrocarbons from depleted reservoirs. This application is essentially an extension of the types of cleaning principles and applications described in the foregoing sections. Surfactant technology has been used in some chemically-enhanced oil recovery (EOR) applications with the objective of mobilizing the residual oil through microemulsion systems that generate ultralow crude oil/water IFT to overcome capillary forces and to enable oil and water to flow [92–94].

To design microemulsion systems for chemical EOR applications that cover a broad range of reservoir temperatures and salinities encountered in formation water (from low to about 30,000 mg/l), formulation studies need to include parameters such as type and temperature, salinity and composition of the crude oil, and concentration of the surfactant. In addition to the phase behavior, IFT and wettability, core flow studies are used to evaluate the microemulsion system in a porous media [94].

Research in chemical EOR started in the 1970s. In the period 2001–2013, the number of publications in this area had significantly increased, as evidence of the large amount of ongoing research at universities and research centers of the oil industry and by surfactant manufacturing companies. However, the recognition of the special properties of microemulsions suggests that development of processes for formation of in-situ microemulsions in depleted petroleum reservoirs is an exciting field for more research and development. In depleted reservoir systems, the natural flow of hydrocarbons from the porous media has effectively ceased, as a result of pressure depletion, as fluid was removed by production of the oil, gas and water. Under such conditions, some hydrocarbon reservoirs may have been subjected to flushing of additional hydrocarbons from the porous media using water-flooding technology. Even if water-flooding and other techniques have been used, there remains some amount of hydrocarbons in the reservoir that are effectively immobile and trapped in dead ends, or tortuous pathways which were bypassed by the path of least resistance physics of the flushing processes.

All of the initial work was directed toward chemical EOR in sandstone reservoirs, because the high-divalent environment encountered in carbonate reservoirs would create salt precipitation problems with the sulfonate surfactants typically used in these applications. Later, the application was extended to carbonate reservoirs with the incorporation of other types of surfactants [93].

At this stage, chemical EOR has advanced to overcome some of the problems encountered in the past. Research at universities and in the petroleum industry has resulted in (1) a reduction of required surfactant concentration, which is economically very important, because continuous injection in reservoirs results in the use of very large volumes of surfactant and other additives; (2) development of a large number of surfactant molecular types to obtain the required phase behavior of the microemulsion system with different crude oils; and (3) better understanding of interactions between the rock formation and alkali/surfactant systems used in chemical EOR injection that has resulted in reduction of surfactant adsorption, lowering the required surfactant concentration.

6.4.8. Microemulsion Cleaning of Fracturing Gels from Shale and Other Rock Formations

A very contemporary application for microemulsion cleaning pertains to development of oil-shale and gas-shale energy resources [95,96]. Shales have the characteristic property of fissility, which enables them to be fractured along planes. When very low porosity shales contain significant amounts of oil and/or gas, one typical method used to produce the hydrocarbons is to fracture the shales using high-pressure injection of fluids that can produce fracture pathways in the shale deposits. Special materials are then injected that prop open the cracks and fissures created by the rock fracturing operation. A gel-type fluid is used to facilitate introduction of proppant materials into the extended fractures. Following the fracturing operation, it is desirable to remove the gel substance to permit easier flow of the hydrocarbons through the fractures into the well. The situation is complicated by the location of the reservoirs far below the surface of the Earth, and by the fact that it is not practical to use mechanical energy to remove the gel material from the fractures. This is a great opportunity for microemulsion technology, because microemulsion solutions can be formulated to incorporate the gel material into the microemulsion structures instantaneously, without need for mechanical energy. To further minimize problems that might arise, if clay minerals are included in the rock, the microemulsion solutions could be formulated using nonaqueous substances, such as CO_2. When the oil or gas wells are placed in production, the pressure of the hydrocarbon fluids can displace the microemulsion solution from the fractures, enabling production enhancement by lowering hydraulic resistance in the fracture pathways and network. The use of microemulsion technology for this challenging application could be very successful.

7. CURRENT TRENDS AND FUTURE DEVELOPMENTS

Microemulsions, incorporating oil and water, have been increasingly used to formulate fluid systems for cleaning applications. Such fluids have the advantage of cleaning with minimal or no requirement for mechanical energy.

Important properties of microemulsions include their ability to form spontaneously, their ability to absorb/solubilize substantial amounts of water and/or oil, their ultralow IFT, and their ability to alter wettability of oil-wetted surfaces. Not surprisingly, microemulsions have found applications in all industries that use cleaning processes. Some current applications include household cleaning (hard and soft surfaces) applications, such as laundry, countertops and floors. Industrial applications include large-scale laundry operations, cleanup of oily wastewater, cleaning organic contaminants on artwork surfaces, such as fresco paintings. Microemulsions have also been used in a number of environmental remediation applications, including solubilization and removal of chlorohydrocarbons from contaminated soil.

The oil and gas industry uses microemulsions in various applications that range from surface cleaning operations (oil removal from drill cuttings and equipment) to downhole well cleaning, and to increase crude oil recovery from rock formations in oil wells. An application where universities and industry have been doing a considerable amount of research is in microemulsion formulations for cleaning reservoir formations to increase oil and gas production.

Another application that started in recent years is the use of microemulsions to increase water flow back in shale gas wells. This application has a good potential for growth, but there are many variables related to fluid/rock interactions and fluid/fluid interactions in the process that need to be defined by experiment and understood.

Microemulsions are quite complex systems that are dependent upon many variables. There remains a wide gap between current scientific knowledge and the practical information available to those who develop fluid formulations for industrial applications.

Future developments of microemulsions for cleaning of surfaces are a wide-open and fertile area for research. Advances will depend upon the advent of a host of new surfactant molecules that can be used to formulate surfactant packages that generate/design microemulsions having capabilities for specific applications, in petroleum recovery, pharmaceuticals, medical practice, chemical reactions, catalysis, decontamination/cleaning of soils and surfaces, cleaning and recovering nanoparticles for reuse, etc. Likewise, future developments may include microemulsion systems that involve pressure-responsive supercritical components [97], such as application in ultradeep wells in crude oil and gas reservoir. Other possible areas of applications may include microemulsions for hydrate control and removal in cold environments subsea and the arctic.

REFERENCES

[1] T.P. Hoar, J.H. Schulman, Transparent water-in-oil dispersions: the oleopathic hydromicelle, Nature 152 (1943) 102.
[2] J.H. Schulman, D.P. Riley, X-ray investigation of the structure of transparent oil–water disperse system, J. Colloid Sci. 3 (1948) 383.
[3] J.-L. Salager, R.E. Anton, D.A. Sabatini, J.H. Harwell, E.J. Acosta, L.I. Tolosa, Enhancing solubilization of microemulsions – state of the art and current trends, J. Surfactants Deterg. 8 (2005) 3.
[4] S. Ezrahi, A. Aserin, N. Garti, Aggregation behaviour in one-phase (Winsor IV) microemulsion systems, in: P. Kumar, K.L. Mittal (Eds.), Handbook of Microemulsion Science and Technology, Marcel Dekker, New York, NY, 1999, pp. 185–244.
[5] J.L. Salager, R. Anton, Ionic microemulsions, in: P. Kumar, K.L. Mittal (Eds.), Handbook of Microemulsion Science and Technology, Marcel Dekker, New York, NY, 1999, pp. 247–280.
[6] M. Xie, X. Zhu, G.W. Miller, D.S. Bohlen, P.K. Vinson, H.T. Davis, L.E. Scriven, Generic patterns in the microstructure of midrange microemulsions, in: S.E. Friberg, B. Lindman (Eds.), Organized Solutions, Surfactant Science Series, vol. 44, Marcel Dekker, New York, NY, 1992, pp. 145–158.
[7] R. Zana, J. Lang, Dynamics of microemulsions, in: S.E. Friberg, P. Bothorel (Eds.), Microemulsions: Structure and Dynamics, CRC Press, Boca Raton, FL, 1987, pp. 153–172.
[8] I. Danielsson, B. Lindman, The definition of microemulsion, Colloids Surf. 3 (1981) 391.
[9] P.A. Winsor, Solvent Properties of Amphiphilic Compounds, Butterworth, London, UK, 1954.
[10] M. Bellocq, Ionic effect of alcohol chain length and salt on phase behavior and critical phenomena in SDS microemulsions, in: P. Kumar, K.L. Mittal (Eds.), Handbook of Microemulsion Science and Technology, Marcel Dekker, New York, NY, 1999, pp. 139–184.
[11] M. Clausse, J. Peyrelasse, J. Heil, C. Boned, B. Lagourette, Bicontinuous structure zones in microemulsions, Nature 293 (1981) 636.
[12] K. Holmberg, Quarter century of progress and new horizons in microemulsions, in: D.O. Shah (Ed.), Micelles, Microemulsions, and Monolayers, Science and Technology, Marcel Dekker, New York, NY, 1998, pp. 161–192.
[13] P.A. Winsor, Binary and multicomponent solutions of amphiphilic compounds. Solubilization and the formation, structure and theoretical significance of liquid crystalline solutions, Chem. Rev. 68 (1968) 1.
[14] J.-L. Salager, J.C. Morgan, R.S. Schechter, W.H. Wade, E. Vasquez, Optimum formulation of surfactant–water–oil systems for minimum interfacial tension or phase behavior, Soc. Petrol. Eng. J. 19 (1979) 107.
[15] M. Bourrel, R.S. Schechter, Microemulsions and Related Systems: Formulation, Solvency, and Physical Properties, Marcel Dekker, New York, NY, 1988.
[16] J.-L. Salager, M. Bourrel, R.S. Schechter, W.H. Wade, Mixing rules for optimum phase behavior formulations of surfactant/oil/water systems, Soc. Petrol. Eng. J. 19 (1979) 271.
[17] M. Bourrel, C. Koukounis, R.S. Schechter, W.H. Wade, Phase and interfacial tension behavior of nonionic surfactants, J. Dispersion Sci. Technol. 1 (1980) 13.
[18] M. Hayes, M. Bourrel, M. El-Emary, R.S. Schechter, W.H. Wade, Interfacial tension and phase behavior of nonionic surfactants, Soc. Petrol. Eng. J. 19 (1979) 349.
[19] M. Bourrel, J.-L. Salager, R.S. Schechter, W.H. Wade, A correlation for phase behavior of nonionic surfactants, J. Colloid Interface Sci. 75 (1980) 451.
[20] J.-L. Salager, Microemulsions, in: G. Broze (Ed.), Handbook of Detergents, Part A: Properties, Surfactant Science Series, vol. 82, Marcel Dekker, New York, NY, 1999, pp. 253–302.

[21] J.-L. Salager, Phase transformation and emulsion inversion on the basis of catastrophe theory, in: P. Becher (Ed.), Encyclopedia of Emulsion Technology, vol. 3, Marcel Dekker, New York, NY, 1988, pp. 79–136.
[22] E.J. Acosta, T. Nguyen, A. Witthayapanyanon, J.H. Harwell, D.A. Sabatini, Linker-biobased bio-compatible microemulsions, Environ. Sci. Technol. 39 (2005) 1275.
[23] J.-L. Salager, R. Anton, J.M. Anderez, J.M. Aubry, Formulation des Microémulsions par la Méthode du HLD, Techniq. de l'Ingénieur (Paris), J2 (2001), pp. 120–157.
[24] A. Witthayapanyanon, J.H. Harwell, D.A. Sabatini, Hydrophilic-lipophilic deviation (HLD) method for characterizing conventional and extended surfactants, J. Colloid Interface Sci. 325 (2008) 259.
[25] K.S. Chan, D.O. Shah, The Effect of Surfactant Partitioning on the Phase Behavior and Phase Inversion of the Middle Phase Microemulsion, Paper SPE 7869, Presented at SPE Oilfield and Geothermal Chemistry Symposium, Houston, TX, 1979.
[26] M. Bourrel, C. Chambu, The rules for achieving high solubilization of brine and oil by amphiphilic molecules, Soc. Petrol. Eng. J. 23 (1983) 327.
[27] M. Miñana-Pérez, A. Graciaa, J. Lachaise, J.-L. Salager, Solubilization of polar oils in microemulsion systems, Progr. Colloid Polym. Sci. 98 (1995) 177.
[28] R. Leung, D.O. Shah, Solubilization and phase equilibria of water–in–oil microemulsions: II. Effects of alcohols, oils, and salinity on single-chain surfactant systems, J. Colloid Interface Sci. 120 (1987) 330.
[29] P.G. de Gennes, C. Taupin, Microemulsions and the flexibility of oil/water interfaces, J. Phys. Chem. 86 (1982) 2294.
[30] L. Quintero, T.A. Jones, G. Pietrangeli, Phase Boundaries of Microemulsion Systems Help to Increase Productivity, Paper SPE 144209, Spe European Formation Damage Conference, Noordwijk, The Netherlands, 2011.
[31] C.A. Miller, Interfacial bending effects and interfacial tensions in microemulsions, J. Dispersion Sci. Technol. 6 (1985) 159.
[32] K. Rakshit, S.P. Moulik, Physicochemistry of W/O microemulsions: formation, stability, and droplet clustering, in: M. Fanun (Ed.), Microemulsions: Properties and Applications, Surfactant Science Series, vol. 144, CRC Press, Boca Raton, FL, 2009, pp. 17–58.
[33] M.J. Hou, M. Kim, D.O. Shah, A light scattering study on the droplet size and interdroplet interaction in microemulsions of AOT–oil–water system, J. Colloid Interface Sci. 123 (1998) 398.
[34] A. Sabatini, E. Acosta, J.H. Harwell, Linker molecules in surfactant mixtures, Curr. Opin. Colloid Interface Sci. 8 (2003) 316.
[35] A. Graciaa, J. Lachaise, C. Cucuphat, M. Bourrel, J.-L. Salager, Improving solubilization in microemulsions with additives. 1. The lipophilic linker role, Langmuir 9 (1993) 669.
[36] A. Graciaa, J. Lachaise, C. Cucuphat, M. Bourrel, J.-L. Salager, Improving solubilization in microemulsions with additives. 2. Long chain alcohols as lipophilic linkers, Langmuir 9 (1993) 3371.
[37] J.-L. Salager, A. Graciaa, J. Lachaise, Improving solubilization in microemulsions with additives. 3. Lipophilic linker optimization, J. Surfactants Deterg. 1 (1998) 403.
[38] H. Uchiyama, E.J. Acosta, S. Tran, D.A. Sabatini, J.H. Harwell, Supersolubilization in chlorinated hydrocarbon microemulsions: solubilization enhancement by lipophilic and hydrophilic linkers, Ind. Eng. Chem. Res. 39 (2000) 2704.
[39] E.J. Acosta, M.A. Le, J.H. Harwell, D.A. Sabatini, Coalescence and solubilization kinetics in linker-modified microemulsions and related systems, Langmuir 19 (2003) 566.

[40] E.J. Acosta, S. Tran, H. Uchiyama, D.A. Sabatini, J.H. Harwell, Formulating chlorinated hydrocarbon microemulsions using linker molecules, Environ. Sci. Technol. 36 (2002) 4618.
[41] E.J. Acosta, P.D. Mai, J.H. Harwell, D.A. Sabatini, Linker modified microemulsions for a variety of oils and surfactants, J. Surfactants Deterg. 6 (2003) 353.
[42] C. Tongcumpou, E.J. Acosta, L.B. Quencer, A.F. Joseph, J.F. Scamehorn, D.A. Sabatini, S. Chavadej, N. Yanumet, Microemulsion formation and detergency with oily soils: I. Phase behavior and interfacial tension, J. Surfactants Deterg. 6 (2003) 191.
[43] J.-L. Salager, C. Scorzza, A. Forgiarini, M.A. Arandia, G. Pietrangeli, L. Manchego, F. Vejar, Amphiphilic Mixtures versus Surfactant Structures with Smooth Polarity Transition Across Interface to Improve Solubilization Performance, Paper O-a17, 7th World Surfactant Congress, CESIO 2008 – Paris, 2008.
[44] G. Goethals, A. Fernandez, P. Martin, M. Miñana-Perez, C. Scorzza, P. Villa, P. Godé, Spacer arm influence on glucidoamphiphile compound properties, Carbohydr. Polym. 45 (2001) 147.
[45] M. Miñana-Pérez, A. Graciaa, J. Lachaise, J.-L. Salager, Solubilization of polar oils with extended surfactants, Colloids Surf. A 100 (1995) 217.
[46] A. Witthayapanyannon, T.T. Phan, T.C. Heitmann, J.H. Harwell, D.A. Sabatini, Interfacial properties of extended-surfactants-based microemulsions and related macroemulsions, J. Surfactants Deterg. 13 (2010) 127.
[47] A. Witthayapanyannon, E.J. Acosta, J.H. Harwell, D.A. Sabatini, Formulation of ultra-low interfacial tension systems using extended surfactants, J. Surfactants Deterg. 9 (2006) 331.
[48] W.H. Wade, J.C. Morgan, J.K. Jacobson, R.S. Schechter, Low interfacial tensions involving mixtures of surfactants, Soc. Petrol. Eng. J. 17 (1977) 122.
[49] R. Antón, F. Mosquera, M. Oduber, Anionic–nonionic surfactant mixture to attain emulsion insensitivity to temperature, Prog. Colloid Polym. Sci. 98 (1995) 85.
[50] M. Bourrel, C. Chambu, R.S. Schechter, W.H. Wade, The topology of phase-boundaries for oil/brine/surfactant systems and its relationship to oil recovery, Soc. Petrol. Eng. J. 22 (1982) 28.
[51] L. Quintero, T.A. Jones, G.A. Pietrangeli, Proper Design Criteria of Microemulsion Treatment Fluids for Enhancing well Production, Paper SPE 154451, SPE Europec/EAGE, Annual Conference, Copenhagen, Denmark, 2012.
[52] C.A. Miller, K.H. Raney, Solubilization–emulsification mechanism of detergency, Colloids Surf. A 74 (1993) 169.
[53] C.A. Miller, Detergency for engineering applications of surfactant solutions, in: P. Somasundaran (Ed.), Encyclopedia of Surface and Colloid Science, Marcel Dekker, New York, NY, 2006, pp. 1664–1669.
[54] P. Tanthakit, S. Chavadej, J.F. Scamehorn, D.A. Sabatini, Ch Tongcumpou, Microemulsion formation and detergency with oily soil: IV. Effect of rinse cycle design, J. Surfactants Deterg. 11 (2008) 117.
[55] Ch Huh, Interfacial tensions and solubilizing ability of a microemulsion phase that coexists with oil and brine, J. Colloid Interface Sci. 79 (1980) 408.
[56] P.G. de Gennes, Wetting: statics and dynamics, Rev. Mod. Phys. 57 (1985) 827.
[57] J.F. Joanny, P.G. de Gennes, A model for contact angle hysteresis, Chem. Phys. 81 (1984) 552.
[58] W. Birch, A. Carré, K.L. Mittal, Wettability techniques to monitor the cleanliness of surfaces, in: R. Kohli, K.L. Mittal (Eds.), Developments in Surface Contamination and Cleaning, William Andrew Publishing, Norwich, NY, 2008, pp. 693–723.
[59] G. Smith, P. Kumar, D. Nguyen, Formulating Cleaning Products with Microemulsions, Paper 164, Proceeding 6th World Surfactant Congress CESIO, Berlin, Germany, 2004.

[60] B. Lindman, U. Olsson, O. Soderman, Characterization of microemulsions, in: P. Kumar, K.L. Mittal (Eds.), Handbook of Microemulsion Science and Technology, Marcel Dekker, New York, NY, 1999, pp. 309–356.

[61] M. Dvolaitsky, M. Guyot, M. Lagues, J.P. Le Pesant, R. Ober, C. Sauterey, C. Taupin, A structural description of liquid particle dispersions: ultracentrifugation and small angle neutron scattering studies of microemulsions, J. Chem. Phys. 69 (1978) 3279.

[62] E. Sjoblom, S. Friberg, Light-scattering and electron microscopy determinations of association structures in W/O microemulsions, J. Colloid Interface Sci. 67 (1978) 16–30.

[63] M. Gasperlin, M. Bester-Rogac, Physicochemical characterization of pharmaceutically applicable microemulsions: Tween40/Imwitor 308/Isopropyl Myristate/Water, in: M. Fanun (Ed.), Microemulsion Properties and Applications, CRC Press, Boca Raton, FL, 2009, pp. 293–309.

[64] J. Drelich, ChFang, C.L. White, Measurement of interfacial tension in fluid–fluid systems, in: A.T. Hubbard (Ed.), Encyclopedia of Surface and Colloid Science, Marcel Dekker, New York, NY, 2002, pp. 3152–3166.

[65] J. Eastoe, Microemulsions, in: T. Cosgrove (Ed.), Colloid Science – Principles, Methods and Applications, Blackwell Publishing, Oxford, UK, 2004, pp. 78–97.

[66] D. Langevin, J. Meunier, Interfacial tension: theory and experiments, in: W.N. Gelbart, A. Ben-Shaul, D. Roux (Eds.), Micelles, Membranes, Microemulsions and Monolayers, Springer, New York, NY, 1994, pp. 485–519.

[67] A. Pandit, C.A. Miller, L. Quintero, Interfacial tensions between bitumen and aqueous surfactant solutions by maximum bubble pressure technique, Colloids Surf. A 98 (1995) 35.

[68] J. Curtis, Environmentally Favorable Terpene Solvents Find Diverse Applications in Stimulation, Sand Control and Cementing Operations, Paper SPE 84124, SPE Annual Technical Conference and Exhibition, Denver, CO, 2003.

[69] J. Robles, M.A. Criado, E. Jensen, W. Morris, Dynamic Mud-Cake Removal Evaluation under Annulus Hydrodynamic Conditions, Paper SPE 95058, SPE Latin American and Caribbean Petroleum Engineering Conference, Rio de Janeiro, Brazil, 2005.

[70] I. A. Frigaarrd, M. Allouche, C. Gabard-Cuoq, Setting Rheological Targets for Chemical Solutions in Mud Removal and Cement Slurry Design, Paper SPE 64998, SPE International Symposium on Oilfield Chemistry, Houston, TX, 2001.

[71] E. Biezen, N. van der Werff, K. Ravi, Experimental and Numerical Study of Drilling Fluid Removal from a Horizontal Wellbore, Paper SPE 62887, SPE Annual Technical Conference and Exhibition, Dallas, TX, 2000.

[72] L. Quintero, Ch. Christian, T. Jones, Mesophase Spacer Designs Raise the Bar for Casing and Riser Clean-Up in Deepwater Applications, Paper SPE 129090, IADC/SPE Drilling Conference and Exhibition, New Orleans, LA, 2010.

[73] L. Quintero, Ch. Christian, B. Halliday, C. White, D. Dean, New Spacer Technology for Cleaning and Water Wetting of Casing and Riser, Paper AADE-08-df-ho-01, AADE Fluids Conference and Exhibition, Houston, TX, 2008.

[74] L. Quintero, J.-L. Salager, A. Forgiarini, G. Pietrangeli, J. Brege, Efficient Displacement of Synthetic or Oil-Based Mud and Transitional Phase Inversion, 1st International Conference on Upstream Engineering and Flow Assurance at 2012 AIChE Spring Meeting, Houston, TX, 2012.

[75] L. Quintero, T. Jones, D. E. Clark, One-Step Acid Removal of an Invert Emulsion, Paper SPE 94604, European Formation Damage Conference, Scheveningen, The Netherlands, 2005.

[76] L. Quintero, T. Jones, D.E. Clark, A. Twynam, Naf Filter Cake Removal Using Microemulsion Technology, Paper SPE 107499, European Formation Damage Conference, Scheveningen, The Netherlands, 2007.

[77] D.B. Bennion, F.B. Thomas, D.W. Bennion, R.F. Bietz, Mechanism of Formation Damage and Permeability Impairment Associated with the Drilling, Completion and Production of Low API Gravity Oil Reservoirs, Paper SPE 30320, International Heavy Oil Symposium, Calgary, Alberta, Canada, 1995.

[78] L. Quintero, T.A. Jones, D.E. Clark, D. Schwertner, Cases History Studies of Production Enhancement in Cased Hole Wells Using Microemulsion Fluids, Paper SPE 121926, European Formation Damage Conference, Scheveningen, The Netherlands, 2009.

[79] S. Pondstabodee, J.F. Scamehorn, S. Chavadej, J.H. Harwell, Cleanup of oily wastewater by froth flotation formation, J. Sep. Sci. Technol. 33 (1998) 591.

[80] S. Watcharasing, W. Kongkowit, S. Chavadej, Motor oil removal from water by continuous froth flotation using extended surfactant: effect of air bubble parameters and surfactant concentration, J. Sep. Purif. Technol. 70 (2009) 179.

[81] J.L. Underwood, K.A. Debelak, D.J. Wilson, Soil cleanup by in-situ surfactant flushing. Viii. Reclamation of multicomponent contaminated sodium dodecylsulfate solutions in surfactant flushing, Sep. Sci. Technol. 30 (1995) 2277.

[82] B.-J. Shiaua, D.A. Sabatinia, J.H. Harwell, Solubilization and microemulsification of chlorinated solvents using direct food additive (edible) surfactants, Ground Water 32 (1994) 561.

[83] J. Childs, E. Acosta, M.D. Annable, M.C. Brooks, C.G. Enfield, J.H. Harwell, M. Hasegawa, R.C. Knox, P.S. Rao, D.A. Sabatini, B. Shiau, E. Szekeres, A.L. Wood, Field demonstration of surfactant-enhanced solubilization of DNAPL at Dover air force base, Delaware, J. Contam. Hydrol. 82 (2006) 1.

[84] B.J. Carroll, Physical aspects of detergency, Colloids Surf. A 74 (1993) 131.

[85] K.H. Raney, W.J. Benton, C.A. Miller, Use of videomicroscopy in diffusion studies of oil–water–surfactant systems, in: D.O. Shah (Ed.), Macro- and Microemulsions, ACS Symp. Ser., 272, American Chemical Society, Washington, D.C, 1985, pp. 193–222.

[86] S. Banerjee, S. Sutanto, J.M. Kleijn, M.J.E. Roosmalen, G.J. Witkamp, M.A. Cohen-Stuart, Colloidal interactions in liquid CO_2 – a dry-cleaning perspective, Adv. Colloid Interface Sci. 175 (2012) 11.

[87] J. Eastoe, C. Yan, A. Mohamed, Microemulsions with CO_2 as a solvent, Curr. Opin. Colloid Interface Sci. 17 (2012) 266.

[88] M.J.E. Van Roosmalen, G.F. Woerlee, G.J. Witkamp, Surfactants for particulate soil removal in dry-cleaning with high-pressure carbon dioxide, J. Supercrit. Fluids 30 (2004) 97.

[89] Z. Guan, J.M. DeSimone, Fluorocarbon-based heterophase polymeric materials: block copolymer surfactants for carbon dioxide applications, Macromolecules 27 (1994) 5527.

[90] E. Carretti, L. Dei, B. Salvadori, P. Baglioni, Microemulsions and micellar solutions for cleaning wall painting surfaces, Stud. Conservation 50 (2005) 128.

[91] E. Carretti, R. Giorgi, D. Berti, P. Baglioni, Oil-in-Water nanocontainers as low environmental impact cleaning tools for works of art: two case studies, Langmuir 23 (2007) 6396.

[92] M. Bourrel, A.M. Lipow, W.H. Wade, R.S. Schechter, J.-L. Salager, Properties of Amphiphile/Oil/Water Systems at an Optimun Formulation for Phase Behavior, Paper SPE 7450, SPE Annual Fall Technical Conference, Houston, TX, 1978.

[93] G.J. Hirasaki, C.A. Miller, M. Puerto, Recent Advances in Surfactant EOR, Paper SPE 115386, SPE Annual Technical Conference and Exhibition, Denver, CO, 2008.

[94] S. Solairaj, Ch. Britton, J. Lu, D.H. Kim, U.P. Weerasooriya, G.A. Pope, New Correlation to Predict the Optimum Surfactant Structure for EOR, Paper SPE 154262, SPE Improved Oil Recovery Symposium, Tulsa, OK, 2012.

[95] D.H. Le, D.S. Dabholkar, J. Mahadevan, K. McQueen, Removal of fracturing gel: a laboratory and modeling investigation accounting for viscous fingering channels, J. Petrol. Sci. Eng. 88–89 (2012) 145.

[96] G. Penny, J.T. Pursley, Field Studies and Completion Fluids to Minimize Damage and Enhance Gas Production in Unconventional Reservoir, Paper SPE 107844, European Formation Damage Conference, Scheveningen, The Netherlands, 2007.

[97] N.F. Carnahan, L. Quintero, On Reversed Micelles, Supercritical Solutions, EOR and Petroleum Reservoirs, Paper SPE 23753, DPE Latin America Petroleum Engineering Conference, Caracas, Venezuela, 1992.

Chapter 3

Dual-Fluid Spray Cleaning Technique for Particle Removal

James T. Snow
DNS Electronics, Carrollton, TX, USA

Masanobu Sato
Dainippon Screen Manufacturing Company, Hikone, Shiga, Japan

Takayoshi Tanaka
Dainippon Screen Manufacturing Company, Hikone, Shiga, Japan

Chapter Outline

1. Introduction	107		5.2.2. Crown Formation	119
2. Particles and Adhesion Forces	108		5.2.3. Impact on Liquid Film	120
3. Cleaning Process Window	108		6. Dual-Fluid Spray Development	122
3.1. Theoretical Predictions	110		7. Advanced Spray Development	126
3.2. Experimental Studies	113		7.1. Nozzle Development	126
4. Overview of Particle Removal Techniques	115		7.2. Droplet Energy Density	127
5. Dual-Fluid Spray Cleaning	116		7.3. Damage Threshold	129
5.1. System Description	116		8. Summary and Prospects	133
5.2. Droplet Impact Energy	117		References	134
5.2.1. Impact on Solid Surface	118			

1. INTRODUCTION

The manufacture of advanced semiconductor devices, whether logic or memory, requires hundreds of separate processing steps with 20–25% of these steps being cleaning-related. The International Technology Roadmap for Semiconductors (ITRS) [1] provides guidelines to help enable the successful manufacture of future devices. As evidenced by the large percentage of cleaning steps in device manufacturing, cleaning is a key process step in the fabrication of semiconductor devices. Cleaning removes surface contaminants, e.g. etch residues, metallics, and particles, from the previous step and prepares the wafer surface for the subsequent process. The removal of particles is a key element of the cleaning step. If the particle size exceeds a certain dimension, referred to as the *critical particle diameter* d_c, it needs to be removed in order to prevent device failure. The critical particle diameter is defined in the ITRS as half of the Metal 1 half-pitch dimension; thus, in 2013 for devices with 20 nm gate lengths (27 nm Metal 1 ½ pitch), the critical particle diameter is 14.2 nm as shown in Table 3.1. This diameter further shrinks to 8.9 nm for 14 nm gate lengths in 2017. The numbers of these sized particles are limited to 13 and 34 particles for 300 and 450 mm wafers, respectively.

Cleaning processes will continuously be challenged because of the further scaling down of device structures and introduction of new materials and three-dimensional device features. Removal of these "killer" particle defects must be carried out without damage to these fragile features and with essentially zero material loss (<0.1 Å) or roughening of exposed surfaces. A variety of techniques have been documented in the literature for particle removal [2]. In this chapter, dual-fluid spray cleaning is presented as it offers high potential for removal of particles from various surfaces.

2. PARTICLES AND ADHESION FORCES

While filters remove particles from the gases and liquids in the wafer environment and the processing fluids, most of the particulate contamination found on wafers originates from prior processing steps, e.g. dry etch residues, chemical mechanical polishing slurries, and photoresist strip residues, or it is generated from mechanically moving parts within the processing equipment. These particulate contaminants are attached to the wafer surface by van der Waals and electrostatic double-layer forces. In addition, the condensation of water between a particle and substrate can also create a large capillary force [3,4]. A comparison of the strength of these forces is provided in Table 3.2 [3].

If the particles are not removed soon after deposition, there can be a decrease in their removal efficiency in the subsequent cleaning steps [5–7]. This aging effect is believed to result from the deformation of the particle causing an increase in adhesion. It has been shown that removal of silicon dioxide (SiO_2) particles under storage conditions of 40% relative humidity is more difficult than comparably sized silicon nitride (Si_3N_4) particles due to presumed formation of silica bridges between the SiO_2 particles and silicon substrate [6]. Aging effects have higher impact on removal of smaller particles compared to larger ones [7].

3. CLEANING PROCESS WINDOW

The amount of force necessary to dislodge a particle must be sufficient to break the adhesion forces binding the particle to the wafer substrate but not too large to cause damage to surrounding structures. This is shown pictorially in Fig. 3.1 [8]. If the energy distribution associated with the applied force is too broad,

TABLE 3.1 ITRS Front End Surface Preparation Roadmap for Critical Particle Size and Number*

Year of Production	2012	2013	2014	2015	2016	2017	2018
DRAM ½ pitch (nm)	31.8	28.3	25.3	22.5	20.0	17.9	15.9
MPU/ASIC M1 ½ pitch (nm)	32	27	24	21	18.9	16.9	15.0
MPU gate length (nm)	22	20	18	17	15.3	14	12.8
Wafer diameter (mm)	300	300	300	450	450	450	450
Critical particle diameter (nm)	15.9	14.2	12.6	11.3	10.0	8.9	8.0
Critical particle count (#/wafer)	12.6	12.6	12.6	34.2	34.2	34.2	34.2
Critical particle count >40 nm (#/wafer)	4.7	2.5	1.5	3.4	3.4	3.4	1.7
Si and oxide loss (Å)/LDD step	0.1	0.1	0.1	0.1	0.1	0.1	0.1

*The data in this table have been extracted from Table FEP11—Front End Surface Preparation Technology Requirements in Ref. [1]. DRAM=dynamic random access memory; MPU=microprocessor unit; ASIC=application-specific integrated circuit; LDD=lightly doped drain.

TABLE 3.2 Forces Acting on 100 nm Particle in Solution

Force	Order of Magnitude (N)	Proportionality (R:Radius)
van der Waals	10^{-7}	R
Electrostatic	10^{-8}	–
Capillary	10^{-8}	R
Drag (water, 10 m/s)	10^{-9}	R

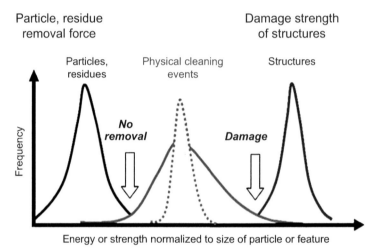

FIGURE 3.1 Cleaning process window depicting particle adhesion force distribution (left) compared with structural integrity force (right) with optimized cleaning forces located in between. A color version of this figure appears in the color plate section.

some particles will not be removed and some damage events may occur; therefore, it is essential that cleaning processes be performed with tightly controlled energies. With the continual shrinking of device dimensions and introduction of new architectures, the strength of these features has also diminished.

3.1. Theoretical Predictions

The forces required to remove a particle have been compared to those that might damage a structure in a semiconductor device both theoretically [3,9–11] and experimentally [12–15]. The capability of different cleaning methods, e.g. standard clean 1 (SC1), megasonics, liquid jet and droplet jet, to remove particles was examined based on particle–substrate interactions [3]. After evaluating the different forces acting on particles (Table 3.2) and the potential differences between

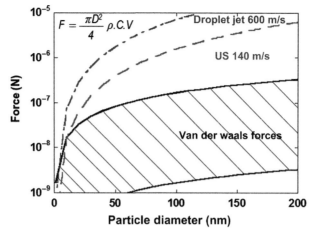

FIGURE 3.2 Droplet jet and ultrasonic (US) shock wave forces exceeding van der Waals particle attachment force.

ideal spherical particles and real-world particles, the researchers selected the system of an ideal aluminum oxide (Al_2O_3) particle on SiO_2 substrate, since this had the largest Hamaker constant (9.6×10^{-20} J in air), and a realistic particle having a van der Waals force two orders of magnitude larger. This range of forces was deemed adequate to cover the majority of forces that might be encountered.

Theoretical performances of cleaning processes were thus examined and the results were grouped into three different types, i.e. universal processes capable of removing all particle types even from patterned wafers, processes similar to the universal group but requiring direct access to the particle, and processes incapable of removing all particles. The processes that involved fast substrate etching for particle lift-off and to enable the electrostatic repulsion to prevent particle redeposition prior to van der Waals reattachment fell into the first group. While this process has no limitation with respect to particle size, the limitations for substrate etching have changed dramatically.

Techniques that generate shock waves, e.g. droplet jet and ultrasonics/megasonics, fell into the second class of processes. The droplet jet process examined had an initial droplet velocity of 400 m/s and calculated speed upon particle impact of ~600 m/s. As shown in Fig. 3.2, the shock wave force F, where D is the particle diameter, ρ is the density of the medium, C is the wave velocity in the medium (1500 m/s in water) and V is the speed of the medium, generated by either the droplet jet or ultrasonics techniques should be capable of removing all particle types provided there was unobstructed access to the particles.

The final group of processes, i.e. those unable to remove all particle types, included methods such as those invoking capillary forces acting on particles at the air/liquid interface of a bath and cleaning techniques using high acceleration, e.g. substrate expansion from laser exposure.

In a subsequent investigation the same researchers expanded the scope to include potential damage to structures, either deposited on the wafer surface or patterned from the substrate (bulk) [10]. In the case of the deposited structures, damage would occur when the external force applied at midheight exceeded the van der Waals force holding the structure to the substrate. The external force was linearly dependent on the line thickness. For the bulk structure, i.e. the structure patterned from the substrate, the damage criterion was selected when the stress at the base of the line exceeded 1 GPa. Calculation of this force was performed by finite element analysis using a computer software program. It was found that the maximum stress component appeared at the center of curvature at the bottom of the line. Calculations were performed for bottom radius r of 1 and 2.5 nm. The collapse forces for the bulk structures were also directly related to the line width for a given aspect ratio, albeit 20 times greater than a deposited structure.

The forces acting on particles and structures, e.g. van der Waals, capillary, drag and shock wave, during various cleaning methods were compared again to determine particle removal vs structure damage. As in the previous study, capillary forces acting on particles at the air/liquid interface of a bath were insufficient to remove all particles. However, for deposited structures with an aspect ratio greater than two, e.g. photoresist lines, it was determined that care should be taken to prevent damage. This damage potential is well known during drying of high-aspect ratio structures and the resulting issues from stiction.

The required flow velocity of 52 m/s for particle removal was found by solving the required drag force necessary to exceed the van der Waals force. While the velocity was independent of particle size, boundary layer effects were not considered. At this velocity, deposited line structures with widths <0.16 µm could be damaged but most structures were thought to be secure.

The shock wave force exerted on a particle or structure is related to the impact area; therefore, higher velocity is needed to remove smaller particles. As shown in Fig. 3.3 [10] for an aspect ratio h/e of 2, the shock wave force needed to remove critical particles can damage patterned lines, even bulk structures at ≤50 nm technology node; however, slight underetching can greatly improve the performance. The researchers concluded that a process using a high-velocity spray has to be carefully tuned in order to remove particles without damage to structures on the wafers.

In a separate study, a numerical simulation using a commercial software package examined the forces involved in particle–substrate interaction in a fluid and the removal mechanism from droplet impact from a dual-fluid spray [11]. Since the dynamic contact angle was included in the operation, the evolution of the droplet impact on the surface and resulting fluid movement could be monitored to evaluate the effectiveness for particle removal. Particles in the size range of 30–100 nm were considered attached to a blanket, unpatterned substrate via van der Waals and electrostatic forces. Deionized water (DIW) was used as the cleaning fluid. Upon droplet impact on the surface, the droplet collapses and the water spreads and flows over the particles. The mechanisms for

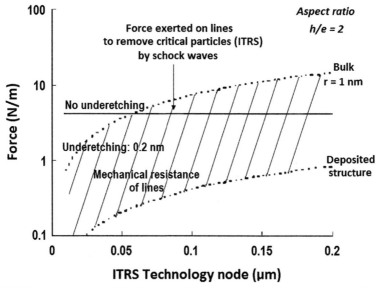

FIGURE 3.3 Comparison of the shock wave drag force necessary to remove all particles with the structural strengths of bulk and deposited structures. *Reproduced by permission of ECS - The Electrochemical Society.*

particle removal were restricted to lift and drag forces caused by the spreading hydrodynamics. Two different cases were computed, i.e. 66 and 100 μm-sized DIW droplets on a hydrophilic surface at an initial droplet velocity of 11.4 m/s.

Results showed that the removal force, whether lift or drag, was similar for both droplet sizes for particles <50 nm. If the particles were larger, the larger sized water droplet provided higher removal force. While the study could not deduce which was the fundamental force for particle removal, it concluded that dual-fluid spray cleaning should be capable of removing 30–100 nm-sized particles.

3.2. Experimental Studies

Several studies have tried to quantify experimentally the removal forces necessary to dislodge particles [12] or to damage Si-based structures [13–15] from silicon wafer surfaces to determine if a cleaning window exists for physical-based cleaning processes (Fig. 3.1). The forces were measured via lateral deflection of an atomic force microscope (AFM) tip as it contacted and dislodged or fractured the deposited particles or silicon structures, respectively, and the resulting angle from this deflection was converted into units of force. The particle removal experiments were performed by depositing either SiO_2 particles or polystyrene latex (PSL) spheres in the range of 100–500 nm onto precleaned, hydrophilic silicon wafers and aging the wafers from 2 to 15 days at 0%, 40% and 100% relative humidity. Using the AFM technique, it was discovered that the silica particles became more difficult to remove as a function of increased

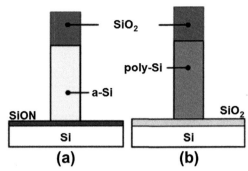

FIGURE 3.4 Test structures of amorphous-Si on SION (a) and polysilicon on SiO_2 (b) surfaces for AFM measurements. © *IOP Publishing 2013. Reproduced by permission of IOP Publishing. All rights reserved.*

aging time and humidity. This was attributed to formation of a water meniscus between the SiO_2 particle and wafer surface since the PSL spheres showed no similar effect. The removal force for the SiO_2 particles varied from ~15 nN for the particles aged for 2 days at 0% RH up to ~500 nN for particles aged for 15 days at 100% RH.

The forces required to damage dummy gate electrode and fin structures have likewise been quantified using the AFM technique. In one study, dummy gate structures were fabricated on the silicon substrate and consisted of layers of either (a) Si-substrate/2.4 nm SiON/100 nm amorphous-Si (a-Si)/60 nm SiO_2 or (b) Si/5 nm SiO_2/100 nm polycrystalline-Si (poly-Si)/60 nm SiO_2 as shown in Fig. 3.4 [13]. Line widths of these gate stacks varied from 80 to 120 nm.

In a separate study [14], dummy 22 nm-wide fins were produced from a-Si deposited on SiO_2 (unannealed or annealed at 1023 K), poly-Si on SiO_2 or crystalline-Si (c-Si) starting from silicon-on-insulator wafers. The height of these fins varied from 60 to 80 nm and spacing between the lines was 1 µm. The same researchers later investigated dummy fins prepared from the same materials but with width of 15 nm and height of 81 nm [15]. The c-Si fins had a width of 20 nm and height of 85 nm. The a-Si fins were cut to different lengths to further investigate the effect of pattern length on collapse force.

Among the fin structures, the a-Si fins required the least force to cause damage (~600 nN), which was attributed to a lower elastic modulus of a-Si compared to poly-Si; c-Si fins required the highest force to damage. Fins with a higher aspect ratio were also damaged more easily. The force F required ranged from 500 to 2000 nN. The collapse forces of the dummy gate stacks were larger, i.e. in the range of 2–9 µN, and varied according to the expression $F = \alpha W^2$, where α is the collapse coefficient and W is the line width.

The forces required to damage these structures were compared with the forces required to remove deposited particles (Fig. 3.5). As shown in Fig. 3.5, a process window theoretically still exists provided the forces utilized in the cleaning methods have a tightly controlled distribution; however, it should be

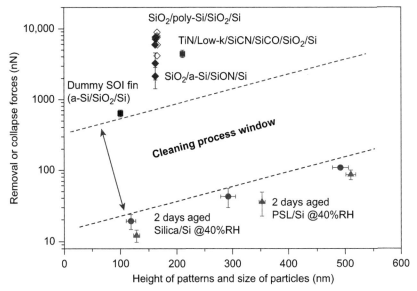

FIGURE 3.5 Cleaning process window showing less energy required to remove particles vs damage structures. Figure provided courtesy of Tae-Gon Kim and used with permission. A color version of this figure appears in the color plate section.

noted that the apparent smaller force required to remove smaller particles compared to larger particles does not take into account boundary layer effects. As technology nodes progress and structures become even more fragile and more prone to damage, it becomes more critical to carefully control the distribution of the particle removal forces.

4. OVERVIEW OF PARTICLE REMOVAL TECHNIQUES

The particle removal techniques can be principally divided into two separate categories, i.e. chemically or physically assisted. Since the introduction of the "RCA Standard Clean" in 1970 [16], ammonia peroxide mixtures (APM) have been used for particle removal, since the high pH (~10) provides a net repulsion of the like-charged particles and the silicon substrate. The particle must first be liberated from the surface via underetching of the silicon substrate, which was easily accomplished with the high concentration of reagents in the original formulation. However, with the reduced substrate loss targets stipulated by the ITRS, this means of particle removal is no longer viable. The use of physical force, coupled when possible with zeta potential effects, is the only avenue currently available.

Physical removal methods most widely reported include systems like megasonics, brush scrubbing, steam-assisted laser cleaning, ice/snow aerosols, and dual-fluid sprays. Among these techniques integrated in single-wafer tools, brush scrubbing provides the highest particle removal efficiency (PRE) [17]; however, due to the forces exerted, it is primarily used for particle removal on the backside of the wafer or the unpatterned front side. Methods using ice/snow aerosols have also been reported [18], but they rely on the aerosol particles being smaller than the patterned structures on the wafer in order to remove particles located between the patterns. The earlier systems had a wide distribution of particles and velocities, which resulted in pattern damage. The technique has been further optimized to provide 10 nm-sized "bullet particles" [19,20], but the cleaning is performed in a vacuum and damage to sensitive structures is not yet known.

In batch-type wafer cleaning tools where the wafers are submerged in a tank filled with a cleaning solution, megasonics has traditionally been employed. The energy released during bubble collapse or cavitation can be adjusted through fine tuning of the appropriate frequency and regulation of the dissolved gas concentration to enable particle removal. However, with highly sensitive structures, it has become more challenging to eliminate the potential to damage the patterns. The demand for single-wafer processing tools exceeded batch tools in 2008 due to better process control, i.e. uniformity, and prevention of cross-contamination. It has been challenging to implement megasonics on a single-wafer platform due to the control that is needed in bubble size, bubble position and control over the sound field, but recent advances have been reported [21,22].

Several early reports compared the performance of megasonics with aerosol spray cleaning [23,24]. In one study, the performance of a mixed-fluid spray jet, whether on a commercial (Nanospray [25]) or experimental system, was compared against various commercial and prototype megasonic tools [23]. The cleaning solutions were either air-saturated DIW or diluted APM solutions, where oxide etch amount was maintained less than 0.5 Å. The removal efficiency of 78 nm SiO_2 particles on hydrophilic 200 mm Si wafers was contrasted with damage to 70 nm poly-Si line structures.

In the case of the mixed-fluid spray cleaning, the PRE and incidence of damage both increased as a function of increased gas flow rate through the nozzle. In addition, with the commercial system, the PRE was highest and consequently more damage events were observed at positions of the wafer where the spray from the nozzle had a higher incidence rate, i.e. at wafer center and edge. The movement of the spray arm could be optimized as shown with the experimental setup to provide a more uniform spray contact time and improve the PRE across the wafer and reduce the amount of center and edge-centric damage sites. At the time this study was conducted, the performance of the mixed-fluid spray jet was superior to any of the evaluated megasonic systems; >80% PRE could be achieved without any damage to the patterned lines.

5. DUAL-FLUID SPRAY CLEANING

5.1. System Description

Dual-fluid sprays are aerosol droplets created from the admixing of a liquid with a gas in a nozzle. The size and velocity distributions of the formed liquid droplets are dependent on the nozzle shape and dimension and the flow rates of the two fluids. Examples of spray nozzles are shown in Fig. 3.6.

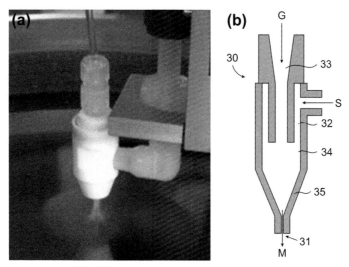

FIGURE 3.6 Examples of spray cleaning nozzles.

A cross-sectional view of a spray nozzle is shown in Fig. 3.6(b) [26,27]. The nozzle has a double pipe structure with the liquid being introduced through an inlet port S and gas through a separate inlet G. The two fluids are mixed in the atomizing zone 34, where a mist consisting of the fluid droplets is formed. There is an acceleration zone 35 downstream of the mixing part 34, which accelerates the formed droplet mist M from the exit orifice 31.

The formed droplets can be characterized by the Weber number W_e given in Eqn (3.1)

$$W_e = \rho v^2 d/\sigma, \qquad (3.1)$$

where ρ is the density of the fluid, v is the velocity, d is the droplet diameter and σ is the surface tension. The formed droplet is accelerated by the gas flow and is inversely related to the droplet diameter according to $1/d\alpha$, where $1<\alpha<1.5$ [28]. The spray technique has been used in various industries for a variety of purposes, e.g. paint on surfaces, insecticides on plants, snow making for winter skiing, cleaning solutions for automobile car washes, to name a few. In addition, the forces of high-speed water droplets on varied surfaces, e.g. jet turbine engines, have been studied in detail to better understand how to mitigate the wear and damage to these surfaces [29].

5.2. Droplet Impact Energy

The resulting effects after a droplet impacts a surface have been widely reported in the literature. Experimental and theoretical studies have been used to better understand the mechanism and generated forces. Studies have been conducted on single droplets impacting solid, dry surfaces [30–36], malleable surfaces

[37,38], droplets impacting a liquid film on solid surface [39], and droplet impact on both solid and liquid surfaces [40–42].

5.2.1. Impact on Solid Surface

During the initial stage of impact on a solid surface, the droplet is misshapen and compacted at its base. The density and compressibility of the liquid, as well as the impact velocity and droplet radius are all important parameters. After impact, the liquid in the droplet adjacent to the contacting surface is extremely compressed, but the remainder of the droplet liquid remains undisturbed.

Early studies of a single droplet impacting a rigid, inelastic surface used a one-dimensional approximation based on the water hammer theory to provide an impact pressure P given by Eqn (3.2):

$$P = \rho_0 C_0 V_0, \qquad (3.2)$$

where P is the generated pressure, ρ_0 is the liquid density, C_0 is the sonic speed of the liquid, and V_0 is the impact velocity. Derivation of this expression assumed that a planar shock wave moved through the liquid at a constant speed C_0. While Eqn (3.2) can be used to describe the pressure created at the moment of first impact, it is insufficient to describe the maximum pressure generated. The assumptions made in the one-dimensional approach were shown by subsequent researchers to be inadequate, e.g. the velocity is not constant and the shape of the shock wave is hemispherical.

The shock wave pressure created after impact is largest at the contact edge to the substrate surface and lowest at the center. There are two different liquid zones formed in the droplet after impact, i.e. compressed and undisturbed, and they are separated by the shock wave front which propagates into the bulk liquid as shown in Fig. 3.7 [32]. The shock wave velocity may be larger than C_0, but for most liquids at low impact velocities the velocities can be assumed to be equal. The generated pressure is also dependent on the target surface, i.e. a more

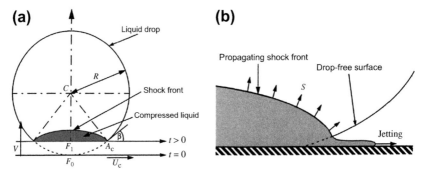

FIGURE 3.7 (a) Liquid drop impact on solid surface showing formation of a shock front. (b) Close-up view moments after impact and start of sideways liquid jetting.

compressible or deformable surface can absorb the energy from the droplet and weaken the resultant forces [39].

A two-dimensional model was later employed to provide a better approximation of the resulting pressures up until the point of droplet spreading [30]. In this analysis, the calculated pressure at the wave front was similar to the result from the one-dimensional approach until the contact angle β in Fig. 3.7 reached half the critical angle, i.e. the radius of the shock wave when it separates from the substrate that was impacted. At this moment, the pressure increased rapidly until approximately three times the water hammer pressure in the impact velocity range of 45–450 m/s.

The hypothesis that a single shock wave is generated was shown by later researchers to result in an anomaly, which could be rectified by proposing a multiple wavelet structure [32]. In this case, the shock wave pressure at the point of contact with the substrate decreased compared to the case when only a single shock wave structure was used.

After droplet impact, the shock wave front is initially pinned to the contacting surface. After only a few nanoseconds when the contact angle exceeds a critical value does the shock wave front separate from the contact edge. At this moment, a sideways jet of liquid is formed as shown in Fig. 3.7(b). The velocity of the formed jet has been experimentally observed to be 2–5 times greater than the initial impact velocity. During this same time, an expansion wave moves into the liquid droplet, which might result in cavitation.

5.2.2. Crown Formation

The impact of a droplet on a shallow liquid layer can exhibit similar behavior as impact on a rough solid surface [43]. Both cases exhibit formation of a crown as shown in Fig. 3.8 [44] with the major difference being the longer time scale in the case of impact on a liquid film. However, if the target surface is a very

FIGURE 3.8 Photograph of crown resulting from droplet impact on liquid surface.

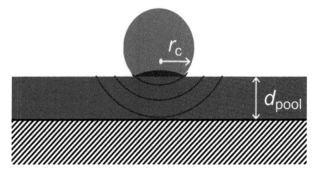

FIGURE 3.9 Droplet impact on liquid layer showing propagation of shock waves.

smooth solid, then a crown is not formed and the liquid simply spreads across the surface. If the frequency of droplet impact is high enough, the surface of the liquid layer may not recover before the arrival of subsequent drops. If the surface is rough, this may affect the impact characteristics of succeeding drops.

Regardless of whether the impact is on a rough solid surface or a liquid film, the velocity of the base of the crown is about 10 times larger than the initial impact velocity. The height of the crown is larger in the case of impact on a liquid layer, since additional liquid is incorporated into the wall of the crown from a small cavity that is formed in the target liquid film. This cavity and crown wall will eventually collapse and form a liquid jet, referred to as a Rayleigh jet, which rises from the center of the former impacted area with the height being dependent on the impact energy.

5.2.3. Impact on Liquid Film

While the splashing characteristics may be similar between droplet impact on a liquid film vs a solid surface, the generated pressure is not. As shown in Fig. 3.9, the shock wave has to traverse the liquid layer to reach the underlying surface [45].

If the radius of droplet contact r_c is small compared to the liquid layer thickness d_{pool} and the thicknesses of the shock wave and liquid layer depth are both constant, the shock wave pressure generated at the substrate surface is given by Eqn (3.3).

$$P_{surface} = \frac{3 p_{wh} \pi r_c^2}{4\pi d_{pool}^2} = \frac{3\rho s v^3 d^2}{8c^2 d_{pool}^2} \tag{3.3}$$

where p_{wh} is the water hammer pressure, r_c is the critical radius (~1 μm), d_{pool} is the depth of the liquid layer, ρ is the liquid density, s is the velocity of the droplet, v is the velocity of the shock wave, d is the diameter of the droplet and c is the speed of sound. Note that Eqn (3.3) includes an additional factor of ½ not included in the original citation, since the water hammer pressure upon impact on water is half of the value compared to the impact on a solid [46].

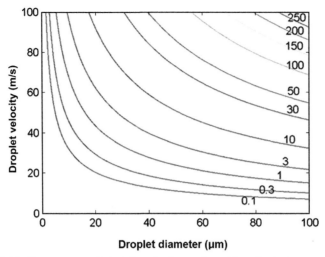

FIGURE 3.10 Pressure curves (in bar) from shock wave formed from droplet impact on a surface coated with a 10 μm liquid film at different velocities with different droplet diameters.

As shown in Eqn (3.3), the pressure upon impact now also depends on the radius of the droplet and the liquid layer thickness. The series of shock wave curves formed from droplet impact on a surface coated with a 10 μm liquid film at different velocities are shown in Fig. 3.10.

After the initial shock wave formation, a crown is formed and the resulting pressure $P_{dynamic}$ on the surface can be estimated by Eqn (3.4).

$$P_{dynamic} = \left[\rho\left(\frac{d\,r_{crown}}{dt}\right)^2 / 2\right]_{t=\frac{d_{pool}}{v}} = \frac{\rho v^2 d^{1.5}}{9.8 d_{pool}^{1.5}} \qquad (3.4)$$

with

$$r_{crown} = \left(\frac{2}{3}\right)^{1/4} \frac{v^{1/2} d^{3/4}}{d_{pool}^{1/4}} t^{1/2}.$$

Note that derivation of this equation neglected any effect from the initial shock wave stage, which should be valid due to the difference in time scales of the two processes. The resulting range of dynamic pressure at the surface of a 10 μm liquid film coated surface at varying droplet sizes and diameters is shown in Fig. 3.11.

As shown, the range of dynamic pressures generated during crown formation is larger than the initial shock wave pressure. Considering the previously reported minimum velocity of 52 m/s necessary for particle removal [10], this corresponds to a minimum dynamic pressure of 13 bar in Fig. 3.11. This value is consistent with the droplet velocity values reported in the literature.

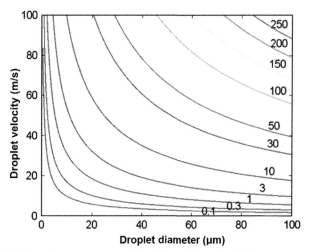

FIGURE 3.11 Dynamic pressure curves (in bar) formed from droplet impact on a surface coated with a 10 μm liquid film at different velocities and droplet diameters.

6. DUAL-FLUID SPRAY DEVELOPMENT

One of the first reports on utilization of the droplet impact technique for particle removal in semiconductor manufacturing was in 1998 [47]. The cleaning performance of a high-velocity water droplet spray was compared against megasonics and high-pressure water jet for removal of PSL particles from a wafer surface. The water droplets in the water droplet spray apparatus, called "M-jet scrubber" had a diameter of 10 μm and were accelerated to the wafer surface with speed varying between 50 and 330 m/s. Particles <1 μm and metal etch residues were removed without damage using the M-jet scrubber. The use of chemical solutions in the M-jet scrubber to improve the removal of small particles was also described. Additional details were subsequently disclosed [48,49].

Other researchers reported on the ability of a supersonic jet cleaning machine to remove 30 nm-sized PSL particles from a 6 inch wafer [50]. This initial report was followed by a study on the cleaning of both the front- and backside of a 200 mm wafer using droplets formed by mixing N_2 gas with DIW at 3.5 kgf/cm^2 (340 kPa) pressure each in a specially designed nozzle [51]. Removal of 0.3 and 0.13 μm-sized particles was compared to roll and disk-type brush scrubbing methods. While roll brush scrubbing provided better PRE, the spray cleaning was slightly better than the disk method. Particle removal was attributed by the researchers to the high impingement pressure from the twin-fluid flow spray and formed side jet of liquid.

Several years later an application of a dual-fluid spray cleaning nozzle for removal of <100 nm particles was reported [52]. The nozzle, called "Soft Spray", dispensed ~20 μm-sized droplets and was installed on a 200 mm single-wafer processing tool. The cleaning effect was termed "physicochemical", i.e.

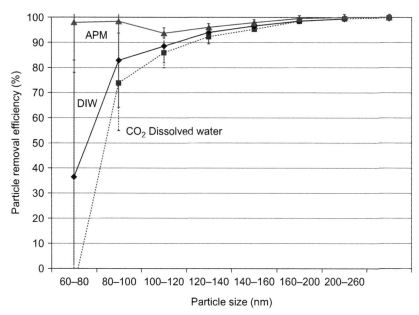

FIGURE 3.12 Averaged PRE for removal of silicon nitride particles for various fluids dispensed through Soft Spray nozzle.

the physical effect from the droplet impact and the chemical effect from the chemistries used as the liquid source. The study evaluated the effects of three different cleaning chemistries, i.e. ozonated water, diluted hydrofluoric acid and APM at three different unspecified concentrations, on wafers that were previously contaminated with either 80 nm monodisperse SiO_2 or polydisperse silicon nitride (Si_3N_4) particles. The oxide etch rate on thermal oxide wafers was also monitored.

The cleaning sequence consisted of a first-step dispensing of hot APM solution followed by spray cleaning with the Soft Spray nozzle. In all cases, complete removal of SiO_2 particles was accomplished regardless of chemistry used in the nozzle. If the Soft Spray step was not included, particle removal was negligible. For removal of nitride particles, the impact of the cleaning chemistry was more evident as shown in Fig. 3.12. Use of APM as the fluid in the spray nozzle provided an averaged removal of >98% of particles >60 nm in size. The improvement in PRE was attributed to better zeta-potential control, i.e. electrostatic repulsion between the Si_3N_4 particles and wafer surface. In addition, the oxide etch rate from the spray step was <1 nm/min; much less than the ~0.14 nm/min etch rate for the hot APM step. The large variation in the data points was due to averaged results from variations in nitrogen (N_2) flow rate and wafer rotation speed.

After these initial publications, several papers appeared in the literature showing the benefits of dual-fluid spray cleaning for particle removal [28,53–57].

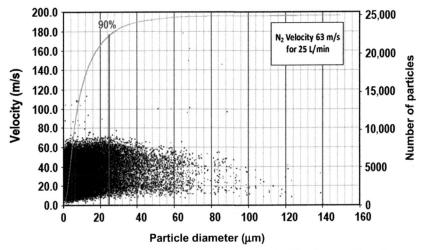

FIGURE 3.13 Droplet size and velocity distribution from atomized liquid spray with gas flow rate of 25 L/min. *Reproduced by permission of ECS - The Electrochemical Society.* A color version of this figure appears in the color plate section.

The effects from changes in the gas and liquid flows on the droplet size and velocity have been described [56,57]. Measurements of individual droplet size and velocity were performed using a Phase Doppler Particle Analysis technique. For an N_2 flow rate of 25 L/min, which resulted in droplet velocity up to 63 m/s, there was a wide distribution of droplet size and velocities as shown in Fig. 3.13.

The arithmetic mean diameter of the 25,000 droplets ranged from 9 to 11 μm depending on the material used to fabricate the nozzle. Ninety percent of the droplets had particle diameters <25 μm. Varying the nitrogen flow rate had an inverse effect on droplet size, e.g. decrease from 30 to 20 standard liters per minute increased the average droplet diameter from 9 to 13 μm.

The effects of fluid flow rates on PRE of 0.1 μm Si_3N_4 particles and damage to 65 nm poly-Si lines were subsequently measured. Increasing the flow rates of DIW and N_2 within the investigated flow ranges improved the PRE. If the N_2 flow velocity was larger than 45 m/s, damage to the poly lines was observed. The influences of the fluid flow rates and nozzle geometry on the formed droplet size and velocity distribution were reported by several researchers subsequently.

As previously shown by Eqns (3.3) and (3.4), the resulting pressures generated upon droplet impact and crown formation are related to the droplet size. With smaller devices on wafers, droplet sizes and velocities also had to decrease to prevent damage. However, generation of a stable droplet distribution is dependent on the size and geometry of the nozzle, so continued optimization of nozzles was required to keep up with the continued introduction of advanced technology nodes.

In 2003 a new nozzle, Nanospray, was introduced that had a different design than the previous Soft Spray nozzle as depicted in Fig. 3.14. The new nozzle was constructed entirely of Teflon, was 3.5 times shorter in length and the

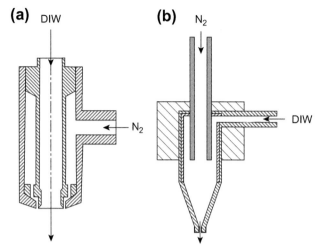

FIGURE 3.14 Schematic drawings of Nanospray (a) and Soft Spray (b) nozzles.

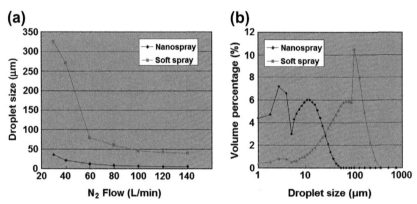

FIGURE 3.15 (a) Effect of nitrogen flow on droplet size for Soft Spray and Nanospray nozzles. (b) Droplet size distribution for Soft Spray and Nanospray nozzles at 80 L/min nitrogen flow.

direction in which the DIW and N_2 fluid flows were introduced into the nozzle were reversed to that used in Soft Spray.

The resulting spray exiting the Nanospray nozzle was more conical in shape; the Soft Spray nozzle delivered a spray that was more columnar. The change in nozzle design provided the desired result of smaller particle diameter with a tighter size distribution. The decrease in droplet size with increase in flow velocity is shown in Fig. 3.15(a), but the change with Nanospray was not as dramatic as with Soft Spray. As shown in Fig. 3.15(b), the averaged droplet size from Nanospray decreased nearly one order of magnitude under conditions of 100 mL/min DIW and 80 L/min of N_2 flows compared to Soft Spray.

Likewise >80% PRE of Si_3N_4 particles (>60 nm) could be achieved at an N_2 flow rate of 65 L/min without damage to 70 nm poly lines as shown in Fig. 3.16.

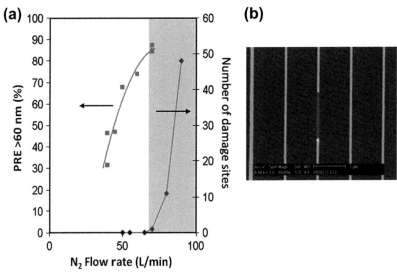

FIGURE 3.16 Effect of N_2 flow on PRE and typical damage to 70 nm poly lines using Nanospray (a). Example of typical damage to 70 nm poly line (b).

In 2006, the Nanospray nozzle was further improved to permit the cleaning of 45 nm lines without damage. To accommodate this, the nozzle was further reduced in size and the Nanospray2 nozzle was introduced to the market [58]. Similar droplet size and PRE could be achieved with Nanospray2 as with its predecessor, but at a reduced gas flow. For example, with the previous Nanospray nozzle, ~50% PRE of Si particles >80 nm could be achieved at an N_2 flow of 30 L/min; however, there were 15 instances of damage to 50 nm poly-Si lines. With Nanospray2 at half the flow rate, the PRE could be improved to 70% with no damage.

As seen in Figs 3.13 and 3.15, there was a large distribution in both the size and velocity of droplets formed by the co-mixing of liquid and gas in a nozzle. For particle removal without damage to sensitive structures as depicted in Fig. 3.1, these distributions had to be tightened to eliminate the presence of outlier droplets of large size causing structural damage. However, droplet creation via the previously described nozzle systems of mixing liquid and gaseous streams is convoluted, i.e. droplet size is inversely related to flow rate of the fluids, in particular the gas flow. A lower flow rate decreases the velocity of the formed droplets, but creates larger droplets which cause more damage. A new approach was required in nozzle design to separate the influence of fluid flow on droplet size.

7. ADVANCED SPRAY DEVELOPMENT

7.1. Nozzle Development

A new nozzle was introduced in 2010 that enabled separate control of droplet diameter and velocity to provide optimized droplet energy [59,60]. The new,

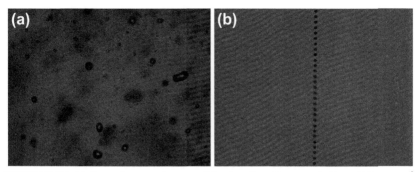

FIGURE 3.17 High-speed camera images of droplets from Nanospray2 (a) and NanosprayÅ (b) nozzles.

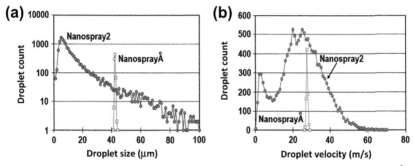

FIGURE 3.18 Droplet size (a) and velocity distributions (b) for Nanospray2 and NanosprayÅ.

specially designed nozzle, called NanosprayÅ, uses a piezoelectric element to discharge evenly sized droplets from many discharge holes in the nozzle. This improved the cleaning efficiency and reduced the potential for pattern damage caused by variations in droplet size and velocity.

The size and velocity of each droplet were verified with a droplet measurement system consisting of a laser, high-speed camera and image processor [61].

Figure 3.17 shows droplets captured by the measurement system [62]. By means of this spray nozzle, the size and velocity of droplets from NanosprayÅ are controlled in the range of ±5% (3σ). The improvement is shown visually in Fig. 3.17 and graphically in Fig. 3.18 when compared to droplets created using the Nanospray2 nozzle. In addition, the biggest advantage of the NanosprayÅ nozzle is that droplet velocity and size can be varied independently while maintaining the uniformity.

7.2. Droplet Energy Density

After verification of uniform droplet size and velocity, the effect on pattern damage was investigated [63]. A NanosprayÅ nozzle was installed on a single-wafer

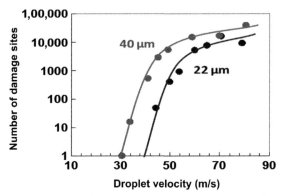

FIGURE 3.19 Damage to photoresist structures for two droplet sizes at varying velocity. A color version of this figure appears in the color plate section.

cleaning tool and spray processing with 22 and 40 µm-sized droplets was followed by spin drying. Three hundred millimeter wafers with photoresist structures (width=250 nm and height=560 nm), which were equivalent to the mechanical strength of 3×nm poly-Si gate structures, were utilized as damage test samples. As droplet velocity increased (Fig. 3.19), the amount of damage increased drastically over the starting threshold.

More damage at a lower threshold velocity was detected in the case of the larger 40 µm-sized droplets. The change in the observed linear relationship between number of damage sites with velocity was ascribed to a measurement irregularity, i.e. the measurement tool recognized damage clusters as a single damage site. Visual inspection with an optical microscope showed a large number of damage sites in the locations marked as one damage event.

The observed effect of droplet size on damage events was explained by examining the kinetic energy of the droplets on the initial impact area. Thus, the kinetic energy E_k of a droplet with diameter d moving at velocity v was calculated from Eqn (3.5).

$$E_k = \frac{\pi \rho d^3 v^2}{12} \quad (3.5)$$

where ρ is the density of liquid. However, the kinetic energy alone was insufficient to explain the observed number of damage events. The droplet energy density E_d expresses the droplet energy within the projected area A of the droplet directly at the beginning of impact. Dividing E_k by A provided E_d as shown in Eqn (3.6).

$$E_d = \frac{E_k}{A} = \frac{\pi \rho d^3 v^2 / 12}{\pi d^2 / 4} = \frac{\rho d v^2}{3} \quad (3.6)$$

This energy E_d was used to explain the correlation of droplet energy to damage occurrence as shown in Fig. 3.20. The E_d for each droplet condition was calculated by substituting experimental values of diameter and velocity.

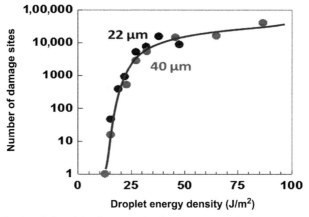

FIGURE 3.20 Correlation of droplet energy density with number of damage sites. A color version of this figure appears in the color plate section.

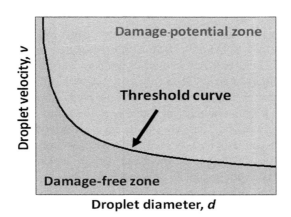

FIGURE 3.21 Threshold energy curve separating damage-free from damage-potential zone.

7.3. Damage Threshold

The concept of damage threshold energy, E_{th}, i.e. when the number of damage sites equals one, was likewise introduced [63]. The relationship between droplet diameter d and velocity v at this threshold condition E_{th} is shown in Eqn (3.7),

$$E_{th} = \frac{\rho d v^2}{3}, \quad \text{therefore} \quad v = \sqrt{\frac{3E_{th}}{\rho d}} \quad (3.7)$$

and could be expressed as a curve separating two zones, damage-free and damage-potential zones, as shown in Fig. 3.21.

If a droplet has E_d less than E_{th}, it is located within the damage-free zone and has no potential for damaging structures. On the other hand, if the value is more than E_{th}, it is located in the zone above the threshold curve and is capable of

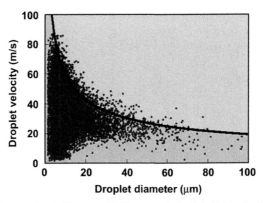

FIGURE 3.22 Damage threshold curve with Nanospray2 droplets. Below the threshold curve is a damage-free zone and above the curve is the damage-potential zone.

causing damage. Of course, this threshold value varies according to the strength of the structure.

For the verification of this threshold curve, droplets dispensed from a Nanospray2 nozzle at 16 L/min N_2 flow were measured and plotted as shown in Fig. 3.22. Three hundred millimeter wafers with photoresist structures (length = 250 nm and height = 560 nm) were again utilized as damage test samples. The number of droplets located in the upper damage-potential zone numbered 500 and a similar number of damages of the photoresist structures was observed experimentally. The cause of damage was ascribed to these outlier droplets, which validated the proposed concept of the threshold energy curve.

These relationships had been applied only to pattern damage generation, and it was not yet known if a similar relationship would be observed with particle removal. In a subsequent paper using a similar approach, the influence of droplet size and velocity for particle removal was investigated with the NanosprayÅ nozzle and contrasted with damage potential [64].

Particle removal experiments were performed on 300 mm wafers, which were previously contaminated with ~10,000 78 nm SiO_2 particles. All processes were conducted on a single-wafer cleaning tool using DIW as the fluid in a NanosprayÅ nozzle. The computer software package STAR-CD was used for simulation of the droplet impact phenomenon on a substrate. The physical strength of poly-Si gate structures of varied widths and SiO_2 particle adhesion forces were measured using the AFM technique as previously described [12–15].

Figure 3.23 shows the relationship between the removed particle counts vs droplet velocity with droplet diameters of 22 μm and 40 μm. It was found that the removed particle counts increased almost linearly with the droplet velocity over the threshold with the larger droplets having a lower threshold velocity. Moreover, it was observed that the larger droplets had a higher PRE at the same velocity.

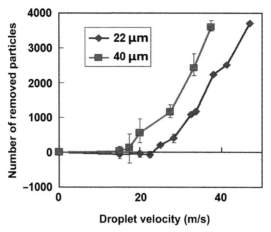

FIGURE 3.23 Relationship between droplet velocity and number of removed 78 nm SiO_2 particles. *Reproduced by permission of ECS - The Electrochemical Society.*

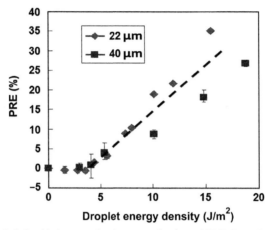

FIGURE 3.24 Relationship between droplet energy density and PRE. *Reproduced by permission of ECS - The Electrochemical Society.*

The droplet energy density E_d from Eqn (3.6) was used to compare the PRE as a function of E_d as shown in Fig. 3.24. As with the pattern damage case, it was found that both droplets had the same threshold energy density E_{th} for particle removal. This threshold value was ~4 J/m², which was smaller than the pattern damage threshold value of ~12 J/m². From this result, it was evident that the theory of droplet energy density could explain the removal of particles, and that the particle removal and pattern damage processes were based on the same mechanism.

While the droplet energy density E_d could be used to explain the results of semiconductor wafer cleaning by dual-fluid spray droplets, the relationship between the physical phenomenon, the droplet impact pressure P and E_d was

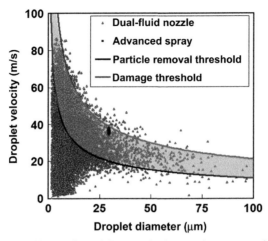

FIGURE 3.25 Threshold curves for particle removal and pattern damage generation with dual-fluid Nanospray2 and NanosprayÅ nozzles. *Reproduced by permission of ECS - The Electrochemical Society.* A color version of this figure appears in the color plate section.

not apparent. For further understanding, the maximum pressure generated by droplet impact on a water-covered substrate was calculated at various velocities and droplet sizes. The water film thickness was held constant at 10 μm and the liquid droplet was assumed to be incompressible.

The results showed that the generated pressure depended on the droplet size d and velocity V^2. The relationships were the same as the relationship of energy density shown in Eqn (3.6). Therefore, the maximum pressure at droplet impact was equivalent to droplet energy density as shown in Eqn (3.8).

$$P \approx E_d = \frac{\rho d V^2}{3} \tag{3.8}$$

The threshold curves for particle removal and pattern damage generation are depicted in Fig. 3.25 along with the distributions of droplet size and velocity for Nanospray2 and NanosprayÅ. The upper curve is the threshold for pattern damage and the lower curve is the threshold for particle removal. The small grouping of points is the distribution of NanosprayÅ spray droplets, while the large distribution of triangles is the droplets from the Nanospray2 nozzle. The gap between the two curves relates to the cleaning process window for particle removal without any pattern collapse. A droplet below this gap has insufficient energy to remove particles and a droplet above this gap is capable of causing pattern damage. To improve cleaning efficiency without pattern damage, all droplets should be controlled in this gap area. The droplets from the Nanospray2 nozzle had a much wider distribution and many droplets existed outside the process window, i.e. ~60% were located below the particle removal threshold curve and ~3% were in the damage-potential zone. In contrast, all droplets from the NanosprayÅ nozzle were situated inside the process window. This result

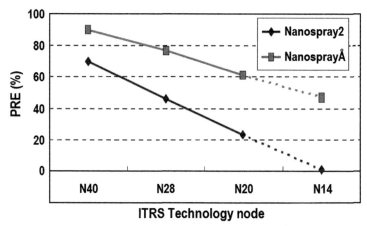

FIGURE 3.26 The PRE performance of Nanospray2 and NanosprayÅ nozzles for removal of PSL spheres.

showed that the droplet size and velocity of the NanosprayÅ droplets could be well-controlled and this technique among dual-fluid aerosol spray methods affords the best performance for particle removal without pattern damage.

The PRE and damage threshold performances with 2×nm technology node devices were evaluated using Nanospray2 and NanosprayÅ nozzles [65]. Using spray settings that would not cause damage, the PRE performance for removal of >45 nm PSL spheres between the two nozzles is shown in Fig. 3.26. The PRE using the NanosprayÅ nozzle was 40% higher than the conventional Nanospray2 nozzle. It was estimated that with 1×nm devices, the conventional dual-fluid spray nozzle will no longer be an effective technique for damage-free particle removal.

8. SUMMARY AND PROSPECTS

The utilization of the forces created from the droplet impact phenomena from dual-fluid spray cleaning has proven to be one of the more effective techniques for particle removal without damage in semiconductor wafer cleaning processing. Further improvement in droplet size and velocity distribution control from the development and introduction of the NanosprayÅ nozzle will extend this technique to future, more sensitive devices. However, particle removal techniques based on fluid flow are more suited for removal of large particles, i.e. smaller particles have a higher adhesion force to the wafer surface with smaller drag force experienced within the fluid boundary layer. It is expected that small particles hidden by high-aspect ratio structures will be impossible to remove with liquid motion-based processes [66].

The use of an interfacial force from freezing of water for particle removal has been reported [67]. A PRE experiment with wafers previously contaminated

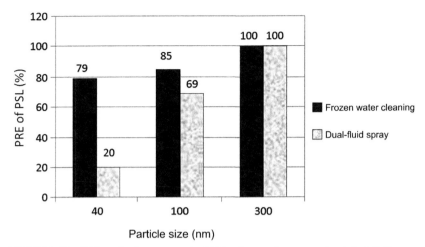

FIGURE 3.27 PRE results from removal of PSL spheres using conventional dual-fluid spray cleaning and frozen water.

with different sizes of PSL particles and damage experiment with wafers patterned with 37 nm poly lines was conducted. Cold water (274 K) was dispensed onto the wafer surface and subsequently frozen by dispensing liquid nitrogen at 83 K. The formed ice was then melted by addition of 353 K hot water and the wafer dried by conventional spin drying. The PRE performance was compared to results obtained using a conventional dual-fluid spray nozzle with N_2 flow of 40 L/min and DIW flow of 100 mL/min as shown in Fig. 3.27.

The dual-fluid spray cleaning settings were known to be damage-free conditions for the 37 nm poly-Si line structures. The frozen water technique also exhibited no damage to the poly-Si lines and had much higher PRE performance, especially with the smallest size particle evaluated. It is expected that the introduction of this new technique will enable removal of small particles for future high technology devices.

REFERENCES

[1] Semiconductor Industry Association. The International Technology Roadmap for Semiconductors, 2012 Update. SEMATECH: Albany, NY, 2012.
[2] K. L. Mittal (Ed.), Particles on Surfaces: Detection, Adhesion and Removal.
    (a) K.L. Mittal (Ed.), Particles on Surfaces: Detection, Adhesion and Removal, vol. 1, Plenum Press, New York, NY, 1988
    (b) K.L. Mittal (Ed.), Particles on Surfaces: Detection, Adhesion and Removal, vol. 2, Plenum Press, New York, NY, 1989
    (c) K.L. Mittal (Ed.), Particles on Surfaces: Detection, Adhesion and Removal, vol. 3, Plenum Press, New York, NY, 1991
    (d) K.L. Mittal (Ed.), Particles on Surfaces: Detection, Adhesion and Removal, vol. 4, Marcel Dekker, New York, NY, 1995

(e) K.L. Mittal (Ed.), Particles on Surfaces: Detection, Adhesion and Removal, vols. 5&6, VSP, Utrecht, The Netherlands, 1999
(f) K.L. Mittal (Ed.), Particles on Surfaces: Detection, Adhesion and Removal, vol. 7, VSP, Utrecht, The Netherlands, 2002
(g) K.L. Mittal (Ed.), Particles on Surfaces: Detection, Adhesion and Removal, vol. 8, VSP, Utrecht, The Netherlands, 2003
(h) K.L. Mittal (Ed.), Particles on Surfaces: Detection, Adhesion and Removal, vol. 9, VSP, Utrecht, The Netherlands, 2006

[3] F. Tardif, A. Danel, O. Raccurt, Understanding of wet and alternative particle removal processes in microelectronics: theoretical capabilities and limitations, Proc. 6th IEEE Symposium on Diagnostics and Yield: Advanced Silicon Devices and Technologies for ULSI Era, Warsaw, Poland, 2003.
[4] R.A. Bowling, A theoretical review of particle adhesion, in: K.L. Mittal (Ed.), Particles on Surfaces: Detection, Adhesion and Removal, vol. 1, Plenum Press, New York, NY, 1988, pp. 129–142.
[5] D.M. Knotter, R. Roucou, R. Peyrin, Reduced particle removal efficiency upon wafer storage, Solid State Phenom. 61 (2009) 145–146.
[6] G. Vereecke, J. Veltens, K. Xu, A. Eitoku, K. Sano, S. Arnauts, K. Kenis, J. Snow, C. Vinckier, P.W. Mertens, Aging phenomena in the removal of nano-particles from Si wafers, Solid State Phenom. 134 (2008) 155.
[7] K. Sano, A. Izumi, A. Eitoku, J. Snow, K. Wostyn, F. Holsteyns, P. Mertens, Single wafer wet cleaning performance and particle removal >36nm, Proc. Sematech Surface Preparation and Cleaning Conference, Austin, Texas, 2006.
[8] P.W. Mertens, G. Vereecke, R. Vos, S. Arnauts, F. Barbagini, T. Bearda, S. Degendt, C. Demaco, A. Eitoku, M. Frank, W. Fyen, L. Hall, D. Hellin, F. Holsteyns, E. Kesters, M. Claes, K. Kim, K. Kenis, H. Kraus, R. Hoyer, T.Q. Le, M. Lux, K.-T. Lee, M. Kocsis, T. Kotani, S. Malhouitre, A. Muscat, B. Onsia, S. Garaud, J. Rip, K. Sano, S. Sioncke, J. Snow, J. Van Hoeymissen, K. Wostyn, K. Xu, V. Parachiev, M. Heyns, Roadblocks and critical aspects of cleaning for sub-65nm technologies, Proc. 2006 International Symposium on VLSI Technology, Systems and Applications, Hsinchu, Taiwan, 2006, pp. 123–126.
[9] N. Hirano, K. Takayama, J. Falcovitz, T. Kataoka, K. Shimada, E. Ando, Microscopic analysis of particle removal by gas/liquid mixture high-speed flow, Solid State Phenom. 207 (1999) 65–66.
[10] F. Tardif, O. Raccurt, J.C. Barbé, F. de Crécy, P. Besson, A. Danel, Mechanical resistance of fine microstructures related to particle cleaning mechanisms, Proc. 8th International Symposium on Cleaning Technology in Semiconductor Device Manufacturing, Electrochemical Society, Pennington, NJ, 2003, pp. 153–160.
[11] Z. Sun, R. Han, Numerical studies on nano-particle removal with micro-droplet spray, Proc. 1st IEEE International Conference on Nano/Micro Engineered and Molecular Systems, Zhuhai, China, 2006, pp. 303–305.
[12] T. Kim, K. Wostyn, T. Bearda, J.G. Park, P. Mertens, M.M. Heyns, Investigation of physical cleaning process window by atomic force microscope, Proc. 11th International Symposium on Cleaning and Surface Conditioning Technology in Semiconductor Device Manufacturing Electrochemical Society, Pennington, NJ, 2009, pp. 203–210.
[13] T.-G. Kim, K. Wostyn, P.W. Mertens, A.A. Busnaina, J.-G. Park, Collapse behavior and forces of multistack nanolines, Nanotechnol 21 (2010) 1.
[14] A. Pacco, T.-G. Kim, P.W. Mertens, Correlation of collapse forces determined by lateral force AFM with damage generation due to physical cleaning processes, Electrochem. Solid State Lett. 14 (2011) H380.

[15] T.-G. Kim, A. Pacco, K. Wostyn, S. Brems, X. Xu, H. Struyf, K. Arstila, B. Vandevelde, J.-G. Park, S. DeGendt, P.W. Mertens, M. Heyns, Effects of interfacial strength and dimension of structures on physical cleaning window, Solid State Phenom. 195 (2012) 123.
[16] W. Kern, D. Puotinen, Cleaning solutions based on hydrogen peroxide for use in silicon semiconductor industry, RCA Rev. 31 (1970) 187.
[17] K. Xu, R. Vos, G. Vereecke, P.W. Mertens, M.M. Heyns, C. Vinckier, J. Fransaer, Mechanisms of particle removal during brush scrubber cleaning, Proc. 8th International Symposium on Cleaning Technology in Semiconductor Device Manufacturing, Electrochemical Society, Pennington, NJ, 2003, pp. 137–144.
[18] S. Banerjee, R. Reidy, L. Rothman, Cryogenic aerosols and supercritical fluid cleaning, in: K. Reinhardt, W. Kern (Eds.), Handbook of Silicon Wafer Cleaning Technology, second ed., William Andrew, Norwich, NY, 2008, pp. 429–478.
[19] K. Hwang, K. Lee, I. Kim, J. Lee, Removal of 10-nm contaminant particles from Si wafers using argon bullet particles, J. Nanoparticle Res. 13 (2011) 4979.
[20] I. Kim, K. Hwang, J. Lee, Removal of 10-nm contaminant particles from Si wafers using CO_2 bullet particles, Nanoscale Res. Lett. 7 (2012) 211.
[21] S. Brems, A. Pacco, H. Struyf, M. Hauptmann, E. Camerotto, P. Mertens, S. De Gendt, Optimizing high frequency ultrasound cleaning in the semiconductor industry, Proc. 8th International Symposium on Cavitation, Singapore, 2012.
[22] M. Hauptmann, F. Frederickx, H. Struyf, P. Mertens, M. Heyns, S. De Gendt, C. Glorieux, S. Brems, Enhancement of cavitation activity and particle removal with pulsed high frequency ultrasound and supersaturation, Ultrason. Sonochem. 20 (2013) 69.
[23] G. Vereecke, T. Veltens, A. Eitoku, K. Sano, G. Doumen, W. Fyen, K. Wostyn, J. Snow, P.W. Mertens, Removal of nano-particles by mixed-fluid jet: evaluation of cleaning performance and comparison with megasonic, Solid State Phenom. 134 (2008) 193.
[24] J. Min, N. Kim, J. Yang, Y. Park, T. Kim, Comparison of jet spray and megasonic module for a cleaning of aluminum layer surface, Proc. 11th International Symposium on Cleaning and Surface Conditioning Technology in Semiconductor Device Manufacturing Electrochemical Society, Pennington, NJ, 2009, pp. 227–232.
[25] Available from Dainippon Screen Manufacturing Co., Ltd., Kyoto, Japan. http://www.screen.co.jp/eng/index.html.
[26] K. Nakajima, M. Sato, H. Sugimoto, A. Hashizume, H. Tsujikawa, Substrate cleaning apparatus and substrate cleaning method, U.S. Patent 7,314,529, 2008.
[27] K. Nakajima, M. Sato, H. Sugimoto, A. Hashizume, H. Tsujikawa, Substrate cleaning apparatus and substrate cleaning method, U.S. Patent 7,422,641, 2008.
[28] Y. Fan, C. Franklin, A. Abit, M. Rouillard, V. Nguyen, T. Krzeminski, E. Brause, A Novel jet spray – meeting the challenge of damage sensitive FEOL cleaning, Proc. Sematech Surface Preparation and Cleaning Conference, Austin, Texas, 2007.
[29] F.P. Bowden, J.E. Field, The brittle fracture of solids by liquid impact, by solid impact, and by shock, Proc. R. Soc. Lond. A 282 (1964) 331.
[30] F.J. Heymann, High-speed impact between a liquid drop and a solid surface, J. Appl. Phys. 40 (1969) 5113.
[31] K.K. Haller, Y. Ventikos, D. Poulikakos, P. Monkewitz, Computational study of high-speed liquid droplet impact, J. Appl. Phys. 92 (2002) 2821.
[32] K.K. Haller, Y. Ventikos, D. Poulikakos, Wave structure in the contact line region during high speed droplet impact on a surface: solution of the Riemann problem for the stiffened gas equation of state, J. Appl. Phys. 93 (2003) 3090.

[33] H.-Y. Kim, S.-Y. Park, K. Min, Imaging the high-speed impact of microdrop on solid surface, Rev. Sci. Instrum. 74 (2003) 4930.
[34] P.A. Lin, A. Ortega, The influence of surface tension and equilibrium contact angle on the spreading and receding of water droplets impacting a solid surface, Proc. 13th IEEE Thermal and Thermomechanical Phenomena in Electronic Systems (ITHERM) Conference, 2012, pp. 1379–1386.
[35] S. Chandra, C.T. Avedisian, On the collision of a droplet with a solid surface, Proc. R. Soc. Lond. A 432 (1991) 13.
[36] M. Bussmann, S. Chandra, J. Mostaghimi, Modeling the splash of a droplet impacting a solid surface, Phys. Fluids 12 (2000) 3121.
[37] K.K. Haller, D. Poulikakos, Y. Ventikos, P. Monkewitz, Shock wave formation in droplet impact on a rigid surface: lateral liquid motion and multiple wave structure in the contact line region, J. Fluid Mech. 490 (2003) 1.
[38] J.E. Field, J.P. Dear, J.E. Ogren, The effects of target compliance on liquid drop impact, J. Appl. Phys. 65 (1989) 533.
[39] T. Sanada, K. Ando, T. Colonius, Effects of target compliance on a high-speed droplet impact, Solid State Phenom. 187 (2012) 137.
[40] F.H. Harlow, J.P. Shannon, Distortion of a splashing liquid drop, Science 157 (1967) 547.
[41] F.H. Harlow, J.P. Shannon, The splash of a liquid drop, J. Appl. Phys. 38 (1967) 3855.
[42] M. Rein, Phenomena of liquid drop impact on solid and liquid surfaces, Fluid Dyn. Res. 12 (1993) 61.
[43] Z. Levin, P.V. Hobbs, Splashing of water drops on solid and wetted surfaces: hydrodynamics and charge separation, Philos. Trans. Royal Soc. Lond. A 269 (1971) 555.
[44] Randy Heisch, Untitled photo, Reprinted with permission. http://photo.net/photodb/photo?photo_id=7209027.
[45] K. Wostyn, M. Wada, K. Sano, M. Andreas, R. Janssens, T. Bearda, L. Leunissen, P. Mertens, Spray systems for cleaning during semiconductor manufacturing, Proc. 22nd European Conference on Liquid Atomization and Spray Systems, Lake Como, Italy, Paper ILASS08-8-2, 2008.
[46] K. Wostyn, Personal Communication, 2012.
[47] I. Kanno, N. Yokoi, K. Sato, Wafer cleaning by water and gas mixture with high velocity, Proc. 5th International Symposium on Cleaning Technology in Semiconductor Device Manufacturing Electrochemical Society, Pennington, NJ, 1998, pp. 54–61.
[48] I. Kanno, Wafer cleaning apparatus, U.S. Patent 5,873,380, 1999.
[49] I. Kanno, M. Tada, M. Ogawa, Two-fluid cleaning jet nozzle and cleaning apparatus, and method utilizing the same, U.S. Patent 5,918,817, 1999.
[50] K. Kitagawa, K. Shimada, K. Nishizaki, Y. Tatehaba, T. Yoneda, E. Andou, Removal of microparticles by supersonic jet cleaning machine, Proc. 58th Autumn Meeting of the Japan Society of Applied Physics and Related Societies, 1997, p. 842.
[51] Y. Tatehaba, K. Kitagawa, K. Shimada, E. Ando, Wafer backside cleaning by twin-fluid flow cleaning, Solid State Phenom. 65-66 (1999) 183–186.
[52] A. Eitoku, J. Snow, R. Vos, M. Sato, S. Hirae, K. Nakajima, M. Nonomura, M. Imai, P.W. Mertens, M.M. Heyns, Removal of small (<100-nm) particles and metal contamination in single-wafer cleaning tool, Solid State Phenom. 92 (2003) 157–160.
[53] Y. Hirota, I. Kanno, K. Fujiwara, H. Nagayasu, S. Shimose, Damage-free wafer cleaning by water and gas mixture jet, Proc. 2005 ISSM, IEEE International Symposium on Semiconductor Manufacturing, 2005, pp. 219–222.

[54] J.M. Lauerhaas, R. Cleavelin, W. Xiong, K. Mochizuki, B. Clappin, T. Schulz, Damage-free cleaning and inspection of advanced multiple-gate FETs, Solid State Technol. 48, March 2008.

[55] K. Xu, S. Pichler, K. Wostyn, G. Cado, C. Springer, G. Gale, M. Dalmer, P.W. Mertens, T. Bearda, E. Gaulhofer, D. Podlesnik, Removal of nano-particles by aerosol spray: effect of droplet size and velocity on cleaning performance, Solid State Phenom. 145–146 (2009) 31.

[56] S. Verhaverbeke, R. Gouk, E. Porras, A. Ko, R. Endo, B. Brown, J.T.C. Lee, An investigation of the critical parameters of an atomized, accelerated liquid spray to remove particles, ECS Trans. 1 (2005) 127.

[57] S. Verhaverbeke, R. Gouk, E. Porras, A. Ko, R. Endo, Using mixed-fluid jet bombardment for advanced particle removal, Solid State Technol. (March 2006) 47.

[58] Dainippon Screen Manufacturing Co., Ltd., Kyoto, Japan. http://www.screen.co.jp/press/pdf/NR060705.pdf.

[59] Establishment of World's First Cleaning Technology for Ultra Miniaturization, Dainippon Screen Manufacturing Co., Ltd., Kyoto, Japan. http://www.screen.co.jp/press/pdf/NR101012E.pdf.

[60] M. Sato, Substrate cleaning method and substrate cleaning apparatus, U.S. Patent Application Publication US2011/0031326, 2011.

[61] Y. Seike, K. Miyachi, T. Shibata, Y. Kobayashi, S. Kurokawa, T. Doi, Silicon wafer cleaning using new liquid aerosol with controlled droplet velocity and size by rotary atomizer method, J. Appl. Phys. 49 (2010) 066701.

[62] Dainippon Screen Manufacturing Co., Ltd., Kyoto, Japan, Product Photos for the Press. http://www.screen.co.jp/eng/press/nr-photo_2009-2011.html.

[63] T. Tanaka, M. Sato, M. Kobayashi, H. Shirakawa, Development of a novel advanced spray technology based on investigation of droplet energy and pattern damage, Solid State Phenom. 187 (2012) 153.

[64] M. Sato, K. Sotoku, K. Yamaguchi, T. Tanaka, M. Kobayashi, S. Nadahara, Analysis on threshold energy of particle removal in spray cleaning technology, Proc. 12th International Symposium on Cleaning and Surface Conditioning Technology in Semiconductor Device Manufacturing, Electrochemical Society, Pennington, NJ, 2011, pp. 75–82.

[65] Y.-H.C. Chien, M. Yeh, S. Ku, C.M. Yang, C.C. Chen, S.M. Jang, K. Izumoto, K. Sotoku, T. Tanaka, M. Sato, H. Shirakawa, M. Tanaka, Physical cleaning enhancement using advanced spray with uniform droplet control, Solid State Phenom. 195 (2013) 195.

[66] S. Brems, M. Hauptmann, Process window – boundary layer effects, Unpublished Report, IMEC, Leuven, Belgium, 2012.

[67] J. Snow and S. Nadahara, Advanced particle removal techniques for <20nm device node, Proc. Sematech Surface Preparation and Cleaning Conference, Austin, TX, 2012.

Chapter 4

Microbial Cleaning for Removal of Surface Contamination

Rajiv Kohli
The Aerospace Corporation, Houston, TX, USA

Chapter Outline

1. Introduction 139
2. Surface Contamination and Cleanliness Levels 140
3. Background 140
   3.1. Microbial Agents 141
4. Principles of Microbial Cleaning 142
5. Cleaning Systems 143
   5.1. Parts Cleaners 144
   5.2. Cleaning Solutions and Microbial Compositions 145
   5.3. Application of Microbes 145
   5.4. Types of Contaminants 145
   5.5. Types of Substrates 146
   5.6. Parts Cleaning 146
      5.6.1. Operating Guidelines 147
   5.7. Costs 148
      5.7.1. Examples of Cost Savings 149
6. Advantages and Disadvantages of Microbial Cleaning 149
   6.1. Advantages 150
   6.2. Disadvantages 150
7. Applications 151
   7.1. Parts Washing 151
   7.2. Oil and Grease Removal 151
   7.3. Sulfate-Reducing Bacteria in Oilfields 152
   7.4. Bacterial Characterization and Monitoring of Surface Cleanliness 153
   7.5. Mercury Bioremediation 153
   7.6. Wound Debridement 154
   7.7. Disinfection and Cleaning 154
   7.8. Cleaning of Historical Art Objects and Structures 155
   7.9. Household and Institutional Applications 156
8. Summary and Conclusions 156
Acknowledgments 157
Disclaimer 157
References 157

1. INTRODUCTION

Solvent cleaning is an established process for removal of surface contaminants in a variety of industrial applications [1]. Many of the conventional solvents used for cleaning, such as hydrochlorofluorocarbons, are considered detrimental to the environment and are increasingly subject to regulations for reduction in their use, and eventual phaseout [2,3]. As a result, there is a continuing effort to find alternate cleaning methods to replace these solvents. One such alternative is microbial cleaning that takes advantage of naturally occurring microbes to remove a wide variety of contaminants from various surfaces. This chapter is focused on the application of microbial cleaning for removal of surface contaminants.

2. SURFACE CONTAMINATION AND CLEANLINESS LEVELS

The most common categories of surface contaminants include particles, thin film or molecular contamination that can be organic or inorganic, ionic contamination, and microbial contamination. Other contaminant categories include metals, toxic and hazardous chemicals, radioactive materials, and biological substances that are identified for surfaces employed in specific industries. Surface contamination can be in many forms and may be present in a variety of states on the surface. Common contamination sources can include machining oils and greases, hydraulic and cleaning fluids, adhesives, waxes, human contamination, and particulates. In addition, a whole host of other chemical contaminants from a variety of sources may soil a surface. Typical cleaning specifications are based on the amount of specific or characteristic contaminant remaining on the surface after it has been cleaned.

Cleanliness levels in precision technology applications are typically specified for particles, as well as for hydrocarbon contamination represented by nonvolatile residue (NVR). For example, civilian and defense space agencies worldwide specify surface cleanliness levels for space hardware for particles in the micrometer per unit area size range and for NVR in the microgram per square centimeter range [4,5]. The cleanliness levels are based on contamination levels established in industry standard IEST-STD-CC1246D for particles from Level 1 to Level 1000 and for NVR from Level AA5 (0.1 ng/cm^2) to Level J (0.025 mg/cm^2) [6]. In many commercial applications, the precision cleanliness level is defined as an organic contaminant level of <10 μg of contaminant/cm^2, although for many applications the requirement is set at 1 μg/cm^2 [6]. These cleanliness levels are either very desirable or are required by the function of parts such as metal devices, electronic assemblies, optical and laser components, precision mechanical parts, and computer parts.

3. BACKGROUND

Microbial cleaning is part of the broader concept of bioremediation. As the name implies, bioremediation is a natural solution to contamination mitigation. It is technically defined as the accelerated breakdown of organic compounds through the use of natural biological agents such as bacteria, enzymes, or fungi. For carbon-based contaminants (grease and oil), the end products are carbon dioxide and water. Bioremediation is a safe, environmentally friendly way to process many kinds of hazardous waste and is supported by the Environmental Protection Agency as a viable solution for cleanup of oil spills and other contaminants, as well as an alternative to solvent cleaning.

3.1. Microbial Agents

There are six main groups of microbes [7].

1. Archaea are a group of unicellular prokaryotic cells that sometimes produce methane during their metabolism. They are specifically adapted to a wide variety of environmental conditions by means of special types of membranes and metabolism.
2. Bacteria are also unicellular prokaryotic organisms. They have a unique type of cell wall and cell membrane that distinguishes them from Archaea. They can digest hydrocarbon contaminants.
3. Fungi are nonphotosynthetic eukaryotes that absorb their nutrients directly from the environment. This group includes mushrooms, molds and yeast.
4. Protista are animal-like, nonphotosynthetic eukaryotes common in moist environments.
5. Viruses are made up of nucleic acids (DNA or RNA) and protein and have some of the characteristics of life. But they lack ribosomes (for protein synthesis), membranes, and means to generate energy, which are properties of cells.
6. Microbial Mergers refer to combinations and collaborations between different microbe species.

Of these microbes, only bacteria (commonly) and fungi (less commonly) have been used for remediation and removal of contaminants [8–14]. When activated, microbes secrete enzymes which break down the contaminants. Hence, pure enzymes manufactured from different microbial strains under aseptic conditions are also used for cleaning [15–17]. Cleaning applications include parts and components cleaning, artworks, oil spills, wastewater, and household and industrial cleaning.

The microbes used in cleaning applications are nonpathogenic and have no recognized hazard potential under ordinary conditions of handling. They are all classified as American Type Culture Collection (ATCC) Class I, are completely safe to humans and the environment, and do not require special biosafety level facilities[1] for handling and use. They are not subject to distribution restrictions by the ATCC, U.S. Department of Health, Public Health Service and the Toxic Substances Control Act (TOSCA).

---

[1] Four biosafety levels have been assigned by the Centers for Disease Control for activities involving microorganisms [23]. The levels are designated in ascending order, by degree of protection provided to personnel, the environment, and the community.

For most surface cleaning applications, the microbes are a highly specialized blend of cultures specifically selected and adapted to degrade a wide range of hydrocarbons. They aggressively attach to and break down oil and grease, but will not attack other substances such as industrial grade metal or natural rubber. The most common strains of bacteria for removal of hydrocarbon contaminants are *Pseudomonas* and *Bacillus* [18–20], while different strains of sulfate-reducing bacteria (SRB) *Desulfovibrio vulgaris* and *Desulfovibrio desulfuricans* are employed for effective cleaning of sulfate contaminants such as calcium sulfate deposits on buildings [21]. In the latter case, the bacteria dissociate calcium sulfate into Ca^{2+} and SO_4^{2-} ions, and further reduce the SO_4^{2-} ion to the S^{2-} ion.

4. PRINCIPLES OF MICROBIAL CLEANING

The basic principle of microbial cleaning for removal of hydrocarbon contaminants involves the reduction of the contaminants into harmless CO_2 and water by the action of microbes [9,22]. Figure 4.1 shows a life cycle diagram of the cleaning process. In a typical surface cleaning application, a cleaning fluid containing a strong surfactant/degreaser contacts the contaminated surface. The surfactant reduces the interfacial tension between the contaminant and the part surface, and separates the contaminants from the surface. A combination of microbes and nutrients is released into (and now living in) the fluid. Nutrients are generally added as part of the cleaning mix to provide emerging microbes with fortification until sufficient amounts of oil and grease have been introduced as carbon sources. The microbes secrete natural enzymes (for example,

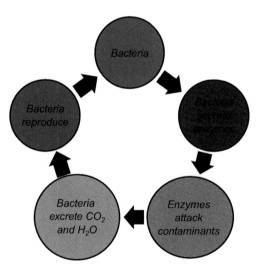

FIGURE 4.1 A typical microbial cleaning life cycle diagram.

lipase [fats, oils], amylase [starches], and protease [proteins]), which cleave the molecular bonds and dissociate the hydrocarbon molecules (contaminants, like oil and grease). This action releases carbon as a source nutrient for the microbes. The microbes are activated and begin to digest the oil and grease that are subsequently absorbed through the cell wall and digested further. The contaminants are then carried by the cleaning fluid through a filtering device, where particulate matter, such as dirt, paint chips, and other items >50 μm, are retained.

In parts cleaners, the cleaning action is due to the surfactant, not the microbes. However, while the microbes do not participate in surface contaminant removal, over time they will remove any hydrocarbons in the cleaning system. In a conducive, nutrient-rich environment, the bioremediation materials continue to manufacture themselves throughout the contaminated solution, increasing the overall biomass of microbes in an exponential manner until all the available hydrocarbons are consumed, thus leaving a clean system with a hydrocarbon-free cleaning solution. Bacteria multiply very rapidly. A single cell can grow to 10^{21} within 24 h [22]. The clean solution can be recirculated through the system, the cleaning cycle is repeated, and there is no interruption in the cleaning process.

Enzymes released by the microbes can only attack one surface of the contaminant. This leads to slower, less effective remediation. The process can be enhanced by a catalyst. Typically, a biocatalyst contains a combination of nonionic surfactants and emulsifiers and water, as well as nutrients that are essential to microbial life. The combination of surfactants and emulsifiers acts to break up the hydrocarbon into very small globules to bring it into intimate contact with the microbes. The globules are surrounded by the enzymes, thereby increasing the rate at which they are dissociated and subsequently digested. The biocatalyst significantly increases bioavailable oxygen. This provides a catalyst for the microbes to multiply faster, resulting in more rapid, more complete bioremediation. The by-products of this process (with pure hydrocarbons) are carbon dioxide, water and soluble fatty acids.

Effective bioremediation systems use a combination of aerobic and anaerobic microorganisms. Aeration provided by the flow of fluid through nozzles and spigots provides adequate additional oxygen to certain strains, while other strains work below the surface in the holding reservoir to break down contaminants that may settle at the bottom of the reservoir.

5. CLEANING SYSTEMS

For parts cleaning applications, the cleaning equipment is especially designed for optimum cleaning performance. Other surface cleaning applications, such as cleaning of artworks and household cleaning, do not require any special equipment.

5.1. Parts Cleaners

Microbial parts cleaners are commercially available in several sizes and models. Figure 4.2 shows examples of free-standing microbial parts cleaning units [24–30]. Typically, these heated cleaning systems consist of an upper sink and a lower tank, filter assemblies to trap visible particulate matter (for example, sand, grit, dirt and paint chips), power module, onboard diagnostics, recirculating pumps, cleaning nozzles, and a tank aeration system that increases the effectiveness of the microbes. Higher pump pressure also improves the cleaning action. The load capacity of these systems ranges from 20 to 200 kg of parts. These units are best for light-duty manual cleaning of parts similar to conventional solvent sink-top units, although recently a larger capacity system

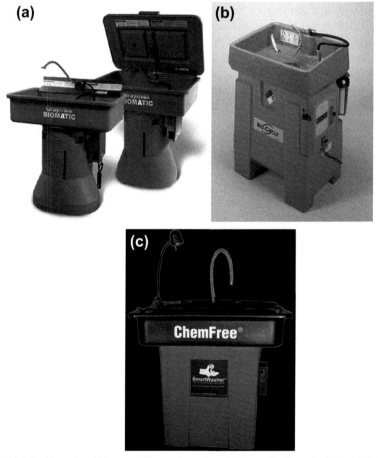

FIGURE 4.2 Examples of bioremediation parts washers. (a) Graymills Biomatics™ Parts Washer, (b) Bio-Circle, and (c) ChemFree SmartWasher [28–30]. *Source: Courtesy of Graymills Corporation, J. Walter Company, and ChemFree Corporation, respectively.*

has been introduced for cleaning entire bicycles [31]. This model includes an integrated bike stand with the parts washer.

5.2. Cleaning Solutions and Microbial Compositions

A wide variety of cleaning solutions and microbial compositions has been developed for many different applications. The powerful degreasing solutions used in cleaning applications are nonhazardous, noncorrosive, pH neutral, nonflammable, nontoxic, noncaustic, aqueous-based degreasing solutions. They are not known to cause damage to humans or the environment. When used in accordance with directions they do not create liquid hazardous wastes or produce cradle-to-grave liabilities. The manufacturers of parts washers offer degreasing solutions that work exclusively in their machines and are not recommended for use in other washers [32–34]. Similarly, the microbial blends are designed for the individual cleaning systems. The specific conditions for elimination of the hydrocarbon contaminants (oil and grease), such as specific temperatures, compensation for foam, aeration parameters, and flow rates, are optimized for the individual units. If the microbial blend is diluted or the cleaning solution composition is changed, it can severely impact the performance of the cleaner. Use of the solutions in other cleaners may affect the digestive effectiveness of the microbes, impair cleaning performance, or even damage the machine, and could void the warranty.

Several manufacturers offer concentrated microbial cleaning solutions that can be used in manual cleaning applications with conventional spray cleaning systems [35–43]. These solutions are used in a typical 20:1 dilution ratio.

Many enzyme-based cleaning compositions have been developed and are available commercially [44–64]. These compositions are formulated from commercially available enzymes [65–67] and are used in varied institutional and household cleaning applications. Section 7 discusses some of these applications.

5.3. Application of Microbes

Industries that perform cleaning of parts prior to rust proofing, phosphating, plating, painting, powder coating, or hot dip galvanizing or coating can benefit from microbial cleaning. Microbes have been successfully used for bioremediation in petrochemical plants, chemical plants, refineries, food processing plants, marine barges, machine shop parts washers, truck washes, wood treating plants and groundwater remediation applications. Examples of applications are discussed in Section 7.

5.4. Types of Contaminants

The cleaning solutions typically contain very strong surfactants, so they will clean a wide variety of contaminants. However, they are designed and are recommended for cleaning biodegradable hydrocarbon contaminants, including the following:

- Crude oil
- Other oils (cutting oil and motor oil)
- Hydraulic transmission fluid
- Solvents
- Btex (benzene, toluene, ethylbenzene, and xylene)
- Greases
- Lubricants
- Amines
- Creosote
- Phenols
- Fats
- Peptide nucleic acid.

The cleaning performance for these types of contaminants is excellent. For example, analyses performed on samples of cleaning solutions from operating bioremediation cleaners have consistently shown hydrocarbon (oil and grease) levels in the 1400 ppm range, compared to an average of 20,000 ppm of oil and grease from nonmicrobial conventional aqueous solvent cleaning [9,11].

Other contaminants that have been successfully treated include paint, ink, glue, adhesive, sealant, wax, tar, graffiti, pen marks, rubber, and resins.

5.5. Types of Substrates

Substrates such as carbon and stainless steels, galvanized steel, brass, copper, aluminum, plastic, and ceramics, fiberglass, glass/quartz, sterling silver, nickel, titanium, plastics, and concrete have been successfully cleaned. Not only is the cleaning solution effective on metal parts, but also will not damage nonmetal components that may be attached to the parts being cleaned such as rubber or plastic fittings. As with all parts cleaners, some surfaces will be cleaned at different rates than others due to the degree and type of contamination on the surface. Because the cleaners operate at a near-neutral pH and lower temperatures, metal parts can be cleaned without etching. Metal, plastic and fiberglass parts will keep their original finish.

5.6. Parts Cleaning

Parts cleaners are simple to operate. As shown in Fig. 4.3, the degreasing solution is sprayed on the contaminated part through the nozzle located in the upper sink. The microbes and nutrients are introduced into the used degreasing solution in the lower tank where the microbes are activated and begin to digest the hydrocarbons in the solution. The clean solution is filtered to trap particulate matter and is recirculated to the upper sink where it can be used to clean additional parts. Heating elements in the lower tank maintain the operating

FIGURE 4.3 Cleaning of parts in a wash basin [30]. *Source: Courtesy of J.Walter Co. Ltd.*

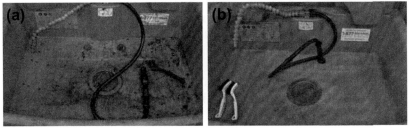

FIGURE 4.4 Parts cleaner sink (a) prior to cleaning, and (b) after cleaning [30]. *Source: Courtesy of J. Walter Co. Ltd.* A color version of this figure appears in the color plate section.

temperature within a range that is ideal for the microbes to thrive, generally 323–360 K. The sink itself is also maintained clean (Fig. 4.4).

In a well-maintained microbial cleaning system, the only regularly generated waste is the used filters that are replaced every 3–8 weeks. The cleaning solution is only replaced when it is no longer effective, which is usually several years. The waste is considered hazardous unless it is tested to demonstrate it is nonhazardous.

5.6.1. Operating Guidelines

Microbial parts cleaning systems are very effective and easy to use. General guidelines will help maintain optimum cleaning performance of the system.

- The cleaning fluid must be heated and aerated constantly to achieve peak cleaning performance. Most microbes require a warm environment to survive and continue to digest the hydrocarbons at an optimal level to clean the solution as quickly as possible. Also, warm solution simply cleans better than cold solution.

- Aggressive chemicals, such as disinfectants, bleach, solvents, acids or chlorinated substances, should not be added to the cleaning solution since they tend to kill the microbes.
- The fluid should be maintained at an optimum level with solutions designed for the unit. If the microbial blend is diluted or the cleaning solution composition is changed, it can severely impact the performance of the system.
- The microbes need time to adapt to the type of contaminants being cleaned. If the microbe solution does not clean effectively at first, cleaning performance will improve after the microbes adapt and digest the new contaminants.
- Very heavily contaminated parts with excessive greases, oils and fluids should be precleaned. Sudden loading of concentrated oils and grease may harm the microbes.
- The filters should be replaced regularly to keep solids from building up at the bottom of the unit and decreasing cleaning performance. The trapped contaminants in the filters can also reach hazardous levels. Replacing the filters can introduce fresh cleaning solution to the existing microbe colony which keeps the system working at an optimum level.
- Parts should be dried after cleaning to prevent rusting or oxidation by residual fluid on the surface. A protective film should be applied to the part before storage.
- Environmental contaminants, such as solvents from aerosols and other sources, can harm microbe populations. Cleaning operations should be performed away from solvent sources.

5.7. Costs

Parts washers are relatively inexpensive, costing around $2500–$8500, depending on the size and capacity of the system [28–31]. Operating costs are generally low. Consumption of cleaning chemicals is minimal since the microbes tend to clean the solution and free up the surfactants to clean and emulsify more contaminants. The premixed or in-situ activated microbial cleaning solution never needs to be replaced, rather, it is topped off in the tank on average every 8–10 weeks to cover losses due to evaporation and fluid left on parts after they are cleaned. The costs of the cleaning solution itself are around $400 for 5 gallons, but it is diluted on average in a 20:1 ratio. Power costs are minimal because there is minimal heat input into the process to maintain an operating temperature in the range 323–360 K. Some system providers offer maintenance contracts at around $600/year for bimonthly service calls [30]. Waste disposal costs for microbial cleaning are low since the primary waste stream is the filters that are replaced every 3–8 weeks. Overall, the costs of microbial cleaning are lower than solvent cleaning, as illustrated by some examples below.

TABLE 4.1 Comparison of the Costs of Conventional Solvent Process with Microbial Cleaning [68]

	Total First-Year Cost	Total Subsequent Yearly Cost
Solvent cleaning	$5050	$3450
Microbial cleaning system	$1820	$1850
Savings	64%	46%

5.7.1. Examples of Cost Savings

The Texas Army National Guard invested approximately $15,000 in August 1995 to purchase 10 bioremediation parts washers to replace solvent cleaners for motor pool operations. In the first year, the Guard eliminated 600 gallons of solvent waste and significantly reduced volatile organic compound emissions, saved $5130 in waste disposal manifesting requirements, and saved $4200 in yearly solvent purchase costs. The estimated payback period was about 18 months. A major aeronautical firm realized savings of more than $80,000 by reducing solvent usage by more than 900 gallons during the first year through use of 23 bioremediation parts washers to replace solvent cleaners [9].

Other studies have shown annual cost savings of nearly 40% by replacing solvent cleaning units with an aqueous cleaner and a microbial cleaning unit with an average payback period of 1.5 years, although in one case, the payback period was <3 months [10,68]. Table 4.1 compares the cost of microbial cleaning with solvent cleaning. The total costs include equipment, cleaning solutions and chemicals, and waste disposal. The subsequent yearly cost for microbial cleaning is slightly higher than the first year which can be attributed to the cost of replenishment of the cleaning solution.

As part of the Lakehurst Pollution Prevention Equipment Program of the U.S. Navy, a solvent-based cleaning system was compared with a bioremediation parts washing system [69,70]. The bioremediation system reduced the waste stream by nearly 100%, saving $1800 in waste disposal costs. In addition, the cleaning solution can be used indefinitely with only occasional replenishment. The equipment is safe to use and does not require personal protective equipment.

6. ADVANTAGES AND DISADVANTAGES OF MICROBIAL CLEANING

The advantages and disadvantages of the microbial cleaning are given in the following sections.

6.1. Advantages

1. This process completely breaks down contaminants to innocuous end products such as water, CO_2, and soluble fatty acids.
2. Microbial cleaning is a natural and safe process. It is a noncorrosive and environmentally friendly cleaning process. No hazardous wastes and emissions are generated.
3. Bioremediation eliminates the need for transportation of spent solvents and other hazardous materials.
4. Microbial cleaning is more economical than traditional solvent cleaning technologies.
5. Cleaning is performed under benign operating conditions with minimal energy input to maintain slightly warmer than ambient temperatures.
6. Microorganisms are nonpathogenic, are completely safe to use and have no recognized hazard potential under ordinary conditions of handling.
7. The rate at which microbes can digest hydrocarbons can approach 80% every 7 days.
8. Most bioremediation parts washers can handle large, tough, dirty jobs.
9. Parts are usually cleaned in the first pass. Even the tiniest crevices and tight spaces in contaminated parts are cleaned because the microbes have close and unhindered contact with the parts.
10. Parts are always exposed to clean solution because the microbes constantly clean the solution and keep the bath clean.
11. Microbes improve the cleaning ability of the cleaning solution. The bioremediation process that takes place in the solution frees the surfactants allowing them to clean and emulsify even more contaminants.
12. Microbes have been successfully used on a variety of contaminants ranging from crude oil to hydrocarbon films.
13. Energy usage is low because of low operating temperature of the process.
14. The process operating costs are low.
15. There is no downtime for maintenance of the system.
16. The cleaning system is simple to use.
17. The costs of waste disposal are low since the filters are the only waste stream generated in low volumes.
18. The cleaning solutions are pH neutral and noncaustic that will not dry, crack or irritate the skin.

6.2. Disadvantages

1. Microbes are susceptible to any biocides designed to kill microbes, such as bleach or strong chemicals that kill living things like some strong pesticides and rat poisons.

2. Added microbes can cohabit with resident bacteria, which can work against the goal of maintaining sanitary conditions in medical and food processing industries, as well as affecting cleaning performance in other applications.
3. The process is limited mainly to removal of biodegradable hydrocarbon contaminants. Most inorganic contaminants, large particles, and other debris cannot be removed.
4. Microbial cleaning is generally not applicable for high-precision cleaning of sensitive parts.
5. The microbial fluid composition is unique to each cleaning system.
6. Filters are the principal waste stream, which must be handled and disposed as hazardous waste.
7. Cleaning may require more scrubbing effort than solvent cleaning.
8. It is difficult to clean heavy or stubborn contaminants.
9. Keeping microbes alive requires proper worker training.
10. Workers may react negatively to certain odors.
11. Cleaning times may be longer than cleaning with conventional solvents.

7. APPLICATIONS

Microbes have been successfully used for remediation in petrochemical plants, chemical plants, refineries, food processing plants, marine barges, truck washes, wood treating plants, oil spill cleanup, soil decontamination, and groundwater remediation applications. Several surface cleaning applications have also been demonstrated including parts washing, oil and grease removal, cleaning of artworks and structures, surface cleaning and disinfection, and household cleaning. The types of contaminants removed include biodegradable oils and greases, lubricants, bacterial contaminants, and sulfate crusts. Some of these cleaning applications are discussed below.

7.1. Parts Washing

This is one of the most common applications for microbial cleaning as discussed at length in Sections 5. Several thousands of parts washers have been installed worldwide and have proven to be cost-effective alternatives to conventional solvent cleaning. In most cases, cleaning effectiveness has been equivalent to, or sometimes even better than, cleaning with solvents.

7.2. Oil and Grease Removal

Industrial activity frequently leaves oil and grease stains on concrete and other floor surfaces, which can build up to a thick layer and can present a safety concern, if it is not removed. Examples are truck bays, machine shop floors, manufacturing facilities, and similar locations. Microbial cleaning has been successfully used to clean up the stains and caked on debris. Figure 4.5 shows a

FIGURE 4.5 Photos of a truck fueling bay before (a) and after (b) microbial cleaning [36]. *Source: Courtesy Worldware Enterprises, Canada.* A color version of this figure appears in the color plate section.

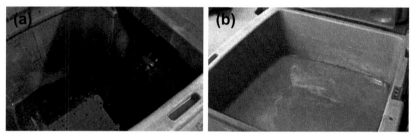

FIGURE 4.6 A cleaning tank (a) after drainage but before cleaning, and (b) after microbial treatment [34]. *Source: Courtesy of J. Walter Co. Ltd.* A color version of this figure appears in the color plate section.

truck fueling bay before and after cleaning with a microbial solution diluted in a 2:1 ratio with water [36]. The solution was sprayed on the contaminated areas (~1670 m^2) and allowed to work for approximately 4 h on the contamination, followed by power washing of the surface. The results are dramatic evidence of the effectiveness of microbial cleaning.

Many examples of heavy oil and grease removal by microbial and enzyme cleaning from drains and grease traps in manufacturing facilities, hospitals, restaurants, food processing facilities, and similar locations have been described on the web sites of the product suppliers [36–43,65–67]. Figure 4.6 shows a cleaning tank heavily contaminated with an oily sludge that was effectively cleaned by microbial solution treatment.

An innovative method of removing oil and grease on slick surfaces is to replace the microbe-enhanced surfactants with protein-enhanced surfactants [12]. This has the benefit that no bacteria are added to the local environment, thus avoiding cohabitation with resident bacteria. The proteins increase the metabolism of the resident bacteria in the wastewater during cleaning or mopping.

7.3. Sulfate-Reducing Bacteria in Oilfields

One of the deleterious consequences of SRB in oilfields is that it can lead to the onset of hydrogen sulfide generation, which can cause corrosion of

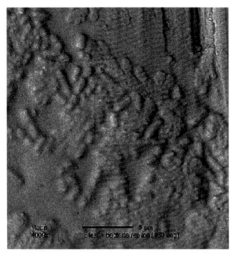

FIGURE 4.7 Inverted image of a cellulose acetate replica of bacteria in a vegetable marinade [74]. The scale bar is 5 μm.

pipelines, platform structures and other equipment, and presents health risks due to the toxicity of H_2S [71–73]. Several microbial processes have been proposed to control SRB contamination in oilfields, including addition of nitrate-reducing bacteria to inhibit H_2S production. These methods have been reviewed recently [73].

7.4. Bacterial Characterization and Monitoring of Surface Cleanliness

Most bacteria are small, approximately 1 μm in diameter, and are not easily removed from a surface. Parts or surfaces cleaned by microbial methods may leave behind bacteria located in scratches, crevices, or similar tight spaces. In-situ visualization and characterization of the bacteria is of interest from both remediation and cleanliness monitoring perspectives. This cannot be done directly because of the large surfaces and fixed installations. Recently, a replication technique using cellulose acetate-replicating tape has been developed to characterize electron microscopically food-borne bacteria on a stainless steel surface [74]. Bacteria are clearly visible in the micrograph of the replica (Fig. 4.7).

Methods for monitoring and measuring the cleanliness of surfaces have been described in detail [75].

7.5. Mercury Bioremediation

Heavy metals, such as mercury, cannot be converted into nontoxic forms by naturally occurring bacteria, but previous attempts have been made to genetically engineer bacteria for heavy-metal remediation without success [76–78].

In a recent study, a transgenic system has been developed for mercury remediation [79]. The proposed system effectively expresses metallothionein (*mt-1*) and polyphosphate kinase (*ppk*) genes in bacteria in order to provide high mercury resistance and accumulation, as high as 80 µM and 120 µM of mercury. This engineered bacterial system presents a viable technology for mercury bioremediation. It may have an application in cleaning mercury-contaminated surfaces.

7.6. Wound Debridement

Several systems have been developed for wound cleansing and debridement using enzyme-based cleaning solutions [80,81 and patents and references cited therein]. Debridement is the surgical excision or enzymatic cleaning of dead, devitalized, or contaminated tissue and removal of foreign matter from a wound to enable healing [82]. These systems are based on the use of pressurized fluid jets to penetrate the skin for delivery and removal of the cleaning solution; a negative-pressure thermotherapeutic fluid delivery device attached to the wound area; pad or dressing with a single or multiple infusion and drainage tubes for continuous delivery of the cleaning solution; or a spray system based on supersonic gas–liquid technology. Most commonly, vegetable-derived proteolytic enzyme solutions are used for cleaning that can include additives, such as activators and inhibitors, to maintain optimum catalytic activity of the enzymes in the cleaning solution [81].

7.7. Disinfection and Cleaning

In the health and food sectors, bacterial and viral infections being transmitted to personnel and patients are a subject of growing concern. One reason for the spread of infection is incomplete or ineffective disinfection of surfaces. Many viruses, bacteria and other pathogens, such as severe acute respiratory syndrome or methicillin-resistant *Staphylococcus aureus* (MRSA), are resistant to existing conventional surface cleaning agents/disinfectants. A new antibacterial cleaning composition has been developed that contains different enzymes (proteolytic, amylolytic, lipolytic, or cellulolytic, or their mixtures) and microbes (*Bacillus* or *Pseudomonas*) together with a surfactant and an aqueous carrier to maintain a minimum 95% catalytic activity at the pH range of 5.5–13.5 [19]. This solution is effective against several resistant bacterial strains, such as MRSA, vancomycin-resistant *Enterococci*, and glycopeptide-intermediate *S. aureus*, and can be used as a cleaning and disinfecting agent in affected areas. It can also be used for killing or inactivating bacteria, viruses, or fungi to prevent spreading of the contaminants. Variations of this composition can be used for cleaning metal, ceramic, glass, or plastic parts, concrete and tile floors, cleaning grease traps, and other household cleaning applications. The contaminants that can be removed include carbon deposits, oil, grease, carbohydrates, starch, and meat and dairy products.

Beyond cleaning and disinfection of surfaces, it is also critical to prevent the growth of microorganisms using a solution such as a quaternary ammonium and benzothiazole composition [83].

Another area of concern is inadequate cleaning and disinfecting of ocular devices such as contact lenses. Several methods have been proposed for cleaning, disinfecting and preserving contact lenses using different microbial cleaning compositions [84,85].

Inadequate cleaning of surgical instruments can result in disastrous consequences for patients in health care facilities [86]. Detergents containing microbial proteases very effectively clean endoscopes and other critical and semicritical surgical instruments. Blood and protein removal during the cleaning is especially critical. Both glutaraldehyde and peracetic acid, used in the disinfection step, are known to fixate residual blood protein. Similarly, removal of body fluids, tissue, residual organic matter, and biofilm is critical to ensure proper cleaning and subsequent high-level disinfection. In general, these detergent formulations offer faster cleaning cycles at lower temperatures, cost savings by extending the lifetime of the instruments, and reduced risk of infections through in-depth cleaning prior to high-level disinfection/sterilization [65].

7.8. Cleaning of Historical Art Objects and Structures

Deterioration of historically and culturally significant monuments, stone structures and artworks is of growing concern [21]. Exposure to the outdoor environment or to uncontrolled indoor environments (temperature and relative humidity) leads to deterioration largely due to atmospheric pollution from a variety of contaminants. Deterioration is a complex process involving chemical, physical, and biological mechanisms. For example, black salt crusts form on stone surfaces as a result of the chemical and microbial interactions between the atmospheric contaminants (sulfur dioxide forming sulfuric acid), the stone (calcium carbonate reacting with sulfuric acid to form calcium sulfate), and microbes that can form calcium oxalate in the crusts. Dust and dirt combine with the calcium sulfate and oxalate, resulting in the black crust. Several microbial techniques have been proposed and successfully demonstrated for cleaning and restoration of stone buildings, frescoes, marble surfaces, and other objects [87–97]. Figure 4.8 shows the *Stories of the Holy Fathers* fresco at the Monumental Cemetery in Pisa, Italy before and after treatment for 2 h with *Pseudomonas stutzeri* bacterial strain [97]. The effects of the treatment on the restoration of the fresco are obvious.

Given the delicate and fragile nature of the surfaces, the cleaning process is almost always a manual process that must be carefully performed. Although biorestoration is promising, the risks posed by the technology have not been sufficiently addressed, as well as the advantages and limitations compared with other physical, chemical and mechanical cleaning processes [21].

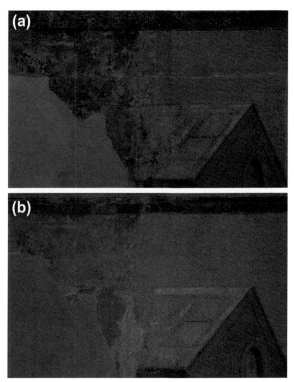

FIGURE 4.8 Effect of biocleaning with *Pseudomonas stutzeri* bacterial strain on the *Stories of the Holy Fathers* fresco before (a) and after (b) treatment [97]. A color version of this figure appears in the color plate section.

7.9. Household and Institutional Applications

One of the most widespread applications of microbial cleaners is as a laundry detergent and for stain and spot removal on fabrics. Other household and institutional applications of microbial cleaning include floors and other hard surfaces in kitchens, bathrooms, locker rooms, garages, loading docks, and similar facilities, tank and equipment cleaning such as ultrafiltration membranes and heat exchangers, as well as for odor control. A wide range of enzymatic formulations have been developed as additives or blends in laundry detergents and other household and institutional applications [44–67,98]. The benefits of microbial cleaning for these applications are effective cleaning performance at lower temperatures, reduced usage of chemicals such as surfactants, increased lifetime of the equipment due to milder cleaning conditions, targeted removal of different contaminants, and lower safety and health risks.

8. SUMMARY AND CONCLUSIONS

Microbial cleaning has been shown to be an effective alternative to conventional solvent cleaning for many applications. The method is based on the affinity of microbes for hydrocarbons which are digested, producing harmless carbon dioxide, water and soluble fatty acids. The microbes are nonpathogenic and safe to handle and dispose. The process is environmentally friendly and is less expensive than solvent cleaning, but it is not applicable to high-precision cleaning applications. Typical applications include parts washing; oil and grease removal from concrete and other floor surfaces, and from drains and grease traps in manufacturing facilities, hospitals, restaurants, food processing facilities, and similar locations; decontamination; cleaning of historical artworks and structures; cleaning and disinfection in health care facilities; wound debridement; controlling SRB in oil fields; mercury bioremediation; and household and institutional cleaning applications.

ACKNOWLEDGMENT

The author would like to thank the members of the STI Library at the Johnson Space Center for help with locating obscure reference articles.

DISCLAIMER

Mention of commercial products in this chapter is for information only and does not imply recommendation or endorsement by The Aerospace Corporation. All trademarks, service marks, and trade names are the property of their respective owners.

REFERENCES

[1] J.B. Durkee, Cleaning with solvents, in: R. Kohli, K.L. Mittal (Eds.), Developments in Surface Contamination and Cleaning, William Andrew Publishing, Norwich, NY, 2008, pp. 759–871. (Chapter 15).
[2] U.S. EPA, The U.S. Solvent Cleaning Industry and the Transition to Non-ozone Depleting Substances, EPA Report U.S. Environmental Protection Agency (EPA), Washington, D.C., 2004. www.epa.gov/ozone/snap/solvents/EPASolventMarketReport.pdf.
[3] U.S. EPA, HCFC Phaseout Schedule, U.S. Environmental Protection Agency, Washington, D.C., 2012. http://www.epa.gov/ozone/title6/phaseout/hcfc.html.
[4] NASA Document JPR 5322.1, Contamination Control Requirements Manual, National Aeronautics and Space Administration, Johnson Space Center, Houston, TX, 2009.
[5] ESA Standard ECSS-Q-70-01B, Space Product Assurance – Cleanliness and Contamination Control, European Space Agency, Noordwijk, The Netherlands, 2008.
[6] IEST Standard IEST-STD-CC1246D, Product Cleanliness Levels and Contamination Control Program, Institute for Environmental Science and Technology (IEST), Rolling Meadows, IL, 2002.
[7] R.M. Atlas, J.C. Philp (Eds.), Bioremediation: Applied Microbial Solutions for Real-World Environmental Cleanup, ASM Press, Washington, D.C., 2005.
[8] R. Dougherty, D. Bassi, Mother nature's wash bath – eliminating drag-out while maintaining clean parts, CleanTech Magazine, April 2004, pp. S9–S11. www.cleantechcentral.com. http://infohouse.p2ric.org/ref/28/27875.pdf.
[9] T.W. McNally, It's alive: letting microbes do the dirty work, Parts Cleaning Magazine, May 1999, pp. 21–27.

[10] Aqueous Parts Cleaning – Best Environmental Practices for Auto Repair, Document 626 Department of Toxic Substances Control (DTSC), California Environmental Protection Agency, Sacramento, CA, 2001. www.dtsc.ca.gov/PollutionPrevention/Vehicle_Service_Repair.html.
[11] O. Ortiz, T.W. McNally, Bioremediation in parts cleaning: fact and fiction, Proceedings CleanTech 2001, Witter Publishing Corporation, Flemington, NJ, 2001, pp. 227–229.
[12] A. Michalow, C. Podella, J. Bladridge, Going green – improved grease and oil cleaning with protein-enhanced surfactants, CleanTech Magazine, June/July 2005, pp. 12–16.
[13] D. Gendel, Bioremediation parts cleaning systems exceed expectations, Process Cleaning Magazine, May/June 2006. http://www.processcleaning.com/articles/bioremediation-parts-cleaning-systems-exceeds-expectations.
[14] H.B. Gunner, M.-J. Coler, W.A. Torello, Antifungal methods, U.S. Patent 7,666,406, 2010.
[15] Enzymes – A Primer on Use and Benefits Today and Tomorrow, White Paper, Enzyme Technical Association, Washington, D.C., 2001.
[16] T. Schäfer, O. Kirk, T.V. Borchert, C.C. Fuglsang, S. Pedersen, S. Salmon, H.S. Olsen, R. Deinhammer, H. Lund, Enzymes for technical applications, in: A. Steinbüchel, S.K. Rhee (Eds.), Polysaccharides and Polyamides in the Food Industry: Properties, Production, and Patents, Wiley-VCH, Weinheim, Germany, 2005, pp. 557–617.
[17] T. Damhus, S. Kaasgaard, H. Lundquist, H.S. Olsen, Enzymes at Work, third ed., Technical Report, Novozymes A/S, Bagsværd, Denmark, 2008. www.novozymes.com.
[18] W.J. Schalitz, J.J. Welch, R.T. Cook, Cleaning composition containing an organic acid and a spore forming microbial composition, U.S. Patent 6,387,874, 2002.
[19] J.T. Manning, Jr., K.T. Anderson, T. Schnell, Enzymatic antibacterial cleaner having high pH stability, U.S. Patent Application 2009/0311136, 2009.
[20] Novo Grease Guard Series Technology, Technical Data Sheet, Novozymes A/S, Bagsværd, Denmark, 2012. www.novozymes.com.
[21] A. Webster, E. May, Bioremediation of weathered-building stone surfaces, Trends Biotechnol. 24 (2006) 255.
[22] Bioremediation Cycle, BESTechnologies Inc., Sarasota, FL, 2010. http://www.bestechcorp.com/bioremediation_cycle.aspx.
[23] Biosafety in Microbiological and Biomedical Laboratories. fifth ed., HHS Publication No. (CDC) 21–1112, Centers for Disease Control and Prevention, U.S. Department of Health and Human Services, Washington, D.C., 2009.
[24] J.L. Strange, Parts washing system, U.S. Patent 6,328,045, 2001.
[25] J.C. McClure, J.L. Strange, Parts washing system, U.S. Patent 6,571,810, 2003.
[26] P.A. Vandenbergh, Bacterial parts washer, composition and method of use, U.S. Patent 6,762,047, 2004.
[27] B.A. Overland, Bioremediation assembly, U.S. Patent 7,303,908, 2007.
[28] SmartWasher Bioremediating Parts Washing System, ChemFree Corporation, Norcross, GA. www.chemfree.com.
[29] Biomatics™ Parts Washers, Graymills Corporation, Chicago, IL. www.graymills.com.
[30] Bio-Circle Parts Cleaning System, J. Walter Co. Ltd, Pointe-Claire, Quebec, Canada. www.biocircle.com.
[31] Smartbike Washer, ChemFree Corporation, Norcross, GA. www.SmartbikeWasher.com.
[32] OzzyJuice®, ChemFree Corporation, Norcross, GA. www.chemfree.com.
[33] Super Biotene Cleaning Solution, Graymills Corporation, Chicago, IL. www.graymills.com.
[34] SC 400 Natural Cleaner/Degreaser, J. Walter Co. Ltd, Pointe-Claire, Quebec, Canada. www.biocircle.com.
[35] ScumBugs Cleaning System, Mineral Masters, West Chicago, IL. www.mineralmasters.com.

Chapter | 4 Microbial Cleaning for Removal of Surface Contamination   159

[36] Eatoils™ Super Degreaser™; Worldware Enterprises Ltd, Cambridge, Ontario, Canada. www.eatoils.com.
[37] Micro-Clean™, Strata International LLC, Glendale, AZ. www.strata-intl.com/Micro-Clean-Oil-And-Gas-Treatments-sc-15.html.
[38] Live Micro 535, Ecoclean Solutions, Farmingdale, NY. www.goecocleansolutions.com.
[39] BioBlitz Products, BESTechnologies, Inc, Sarasota, FL. www.bestechcorp.com.
[40] BioRem 2000 Surface Cleaner and KAS Parts Cleaner Liquid, Technical Data Sheet, Infinite Green Solutions, Phoenix, AZ. http://cleangreenworld.com.
[41] Industrial Enzyme Cleaner and Degreaser, ArroChem Incorporated, Mt. Holly, NC. www.arrochem.com.
[42] Tergazyme Enzyme-Active Powdered Detergent, Technical Bulletin, Alconox, Inc., White Plains, NY. www.alconox.com.
[43] WonderMicrobes, WonderChem Incorporated, Woodburn, KY. www.microbes.wonderchem.com.
[44] T. Cayle, Stabilized aqueous enzyme solutions, U.S. Patent 3,296,094, 1967.
[45] M.M. Weber, Liquid cleaning composition containing stabilized enzymes, U.S. Patent 4,169,817, 1979.
[46] B.J. Anderson, Enzyme Detergent Composition, U.S. Patent 4,404,128, 1983.
[47] R.O. Richardson, A.F. Bromirski, L.T. Davis, Liquid cleaner containing viable microorganisms, U.S. Patent 4,655,794, 1987.
[48] D.A. Estell, Liquid detergent with stabilized enzyme, U.S. Patent 5,178,789, 1993.
[49] L.J. Guinn, J.L. Smith, Microbial cleaner, U.S. Patent 5,364,789, 1994.
[50] W.M. Griffin, R.T. Ritter, D.A. Dent, Drain opener formulation, U.S. Patent 5,449,619, 1995.
[51] Y. Miyota, S. Fukuyama, T. Yoneda, Alkaline protease, process for the production thereof, use thereof, and microorganism producing the same, WIPO Patent WO97/16541, 1997 (in Japanese).
[52] H.A. Nair, G.G. Staud, J.M. Velazquez, Thickened, highly aqueous, cost effective liquid detergent compositions, U.S. Patent 5,731,278, 1998.
[53] D.A. Ihns, W. Schmidt, F.R. Richter, Proteolytic enzyme cleaner, U.S. Patent 5,861,366, 1999.
[54] P.A. Vandenbergh, B.S. Kunka, H.K. Trivedi, Storage stable pseudomonas compositions and method of use thereof, U.S. Patent 5,980,747, 1999.
[55] P.N. Christensen, B. Kalum, O. Andresen, Detergent composition comprising a glycolipid and anionic surfactant for cleaning hard surfaces, U.S. Patent 5,998,344, 1999.
[56] W.J. Schalitz, J.J. Welch, T.R. Cook, Aqueous disinfectant and hard surface cleaning composition and method of use, U.S. Patent 6,165,965, 2000.
[57] C.L. Wiatr, D. Elliott, Composition and methods for cleaning surfaces, U.S. Patent 6,080,244, 2000.
[58] M.E. Besse, R.O. Ruhr, G.K. Wichmann, T.A. Gutzmann, Thickened hard surface cleaner, U.S. Patent 6,268,324, 2001.
[59] D.C. Sutton, Surface maintenance composition, U.S. Patent 6,635,609, 2003.
[60] K.J. Molinaro, D.E. Pedersen, J.P. Magnuson, M.E. Besse, J. Steep, V.F. Man, Stable antimicrobial compositions including spore, bacteria, fungi, and/or enzyme, U.S. Patent 7,795,199, 2010.
[61] Biokleen All Purpose Cleaner, Bi-O-Kleen Industries, Inc., Vancouver, WA. http://biokleenhome.com/products/pro/general.
[62] PSF 110 Natural Enzyme Sport Surface Cleaner, Professional Sports Field Services, LLC, McComb, OH. www.psfs.us.
[63] Drano Max Build-Up Remover, S.C. Johnson, Racine, WI. www.scjohnson.com.

[64] Enzyme Magic Household Products, Enzyme Solutions Incorporated, Garrett, IN. www.enzymesolutions.com.
[65] Novozymes Superior, Sustainable I&I Cleaning Solutions, Novozymes A/S, Bagsværd, Denmark. www.novozymes.com.
[66] SEBrite MI Liquid and Powder, Specialty Enzymes & Biotechnologies, Chino, CA. www.specialtyenzymes.com.
[67] Biogrease GDS, Product Information Sheet, Enzyme Supplies Limited, Oxford, UK. www.enzymesupplies.com/Biogrease_GDSpdf.pdf.
[68] Bio-Circle Parts Cleaning – A Cost-Effective Solution, J. Walter Co. Ltd, Pointe-claire, Quebec, Canada. www.biocircle.com/en-ca/low-cost.
[69] Navy PPEP Pollution Prevention Equipment Program Book. U.S. Department of Defense, Washington, D.C., 2001. http://infohouse.p2ric.org/ref/20/19926/PPEP/PPEPBook.html.
[70] D. Makaruk, A. Caplan, Parts washing using bioremediation technology, Commercial Technologies for Maintenance Activities (CTMA) Symposium 2007, San Antonio, TX, March 28, 2007.
[71] R. Cord-Ruwisch, W. Kleinitz, F. Widdel, Sulfatreduzierende Bakterien in einem Erdölfeld – Arten und Wachstumsbedingungen, Erdöl Erdgas Kohle 102 (1986) 281.
[72] R. Cord-Ruwisch, W. Kleinitz, F. Widdel, Sulfate-reducing bacteria and their activities in oil production, J. Petrol. Technol. 39 (1987) 97.
[73] N. Youssef, M.S. Elshahed, M.J. McInerney, Microbial processes in oil fields: culprits, problems, and opportunities, Adv. Appl. Microbiol. 66 (2009) 141.
[74] M. Kaláb, Replication and Scanning Electron Microscopy of Metal Surfaces Used in Food Processing, White Paper, 2005. http://www.magma.ca/~scimat/Replication.htm.
[75] R. Kohli, Methods for monitoring and measuring cleanliness of surfaces, in: R. Kohli, K.L. Mittal (Eds.), Developments in Surface Contamination and Cleaning, vol. 4, Elsevier, Oxford, UK, 2012, pp. 107–178. (Chapter 3).
[76] S. Chen, E. Kim, M.L. Shuler, D.B. Wilson, Hg2+ removal by genetically engineered *Escherichia coli* in a hollow fiber bioreactor, Biotechnol. Prog. 14 (1998) 667.
[77] M. Valls, V. de Lorenzo, Exploiting the genetic and biochemical capacities of bacteria for the remediation of heavy metal pollution, FEMS Microbiol. Rev. 26 (2002) 327.
[78] J.D. Park, Y. Liu, C.D. Klaassen, Protective effect of metallothionein against the toxicity of cadmium and other metals, Toxicology 163 (2001) 93.
[79] O. N Ruiz, D. Alvarez, G. Gonzalez-Ruiz, C. Torres, Characterization of mercury bioremediation by transgenic bacteria expressing metallothionein and polyphosphate kinase, BMC Biotechnol. 11 (2011) 82.
[80] R. Kohli, Alternate semi-aqueous precision cleaning techniques: steam cleaning and supersonic gas/liquid cleaning systems, in: R. Kohli, K.L. Mittal (Eds.), Developments in Surface Contamination and Cleaning, vol. 3, Elsevier, Oxford, UK, 2012, pp. 201–237. (Chapter 6).
[81] A. Freeman, E. Hirszowicz, M. Be'eri-Lipperman, Apparatus and methods for enzymatic debridement of skin Lesions, U.S. Patent 8,128,589, 2012.
[82] American Heritage Medical Dictionary, Houghton Mifflin Company, New York, NY, 2007.
[83] Y. Ito, Y. Nomura, S. Katayama, Quaternary ammonium and benzothiazole microbicidal preservative composition, U.S. Patent 4,839,373, 1989.
[84] L. Ogunbiyi, T.M. Riedhammer, F.X. Smith, Method for enzymatic cleaning and disinfecting contact lenses, U.S. Patent 4,614,549, 1986.
[85] A. Nakagawa, Y. Oi, Method for cleaning, preserving and disinfecting contact lenses, U.S. Patent 5,409,546, 1998.

[86] N. Chobin, Providing safe surgical instruments: factors to consider, Technical Paper, Infect. Control Today, April 2008. www.infectioncontroltoday.com.
[87] B. von Gilsa, Gemäldereinigung mit Enzymen, Harzseifen und Emulsionen, Zeit. Kunsttechnologie Konservierung 5 (1991) 48.
[88] K.L. Gauri, L. Parks, J. Jaynes, R. Atlas, Removal of sulphated-crust from marble using sulphate-reducing bacteria, in: R.G.M. Webster (Ed.), Proceedings International Conference on Stone Cleaning and the Nature, Soiling and Decay Mechanisms of Stone, Donhead, London, UK, 1992, pp. 160–165.
[89] G. Ranalli, M. Chiavarini, V. Guidetti, F. Marsala, M. Matteini, E. Zanardini, C. Sorlini, The use of microorganisms for the removal of sulphates on artistic stoneworks, Int. Biodeterior. Biodegradation 40 (1997) 255.
[90] G. Ranalli, G. Alfano, C. Belli, G. Lustrato, M.P. Colombini, I. Bonaduce, E. Zanardini, P. Abbruscato, F. Cappitelli, C. Sorlini, Biotechnology applied to cultural heritage: biorestoration of frescoes using viable bacterial cells and enzymes, J. Appl. Microbiol. 98 (2005) 73.
[91] C. Todaro, Gil enzimi: limiti e potenzialità nel campo della pulitura delle pitture murali, in: G. Biscontin, G. Driussi (Eds.), Proc. XXI International Congress Scienza e Beni Culturali: Sulle Pitture Murali. Riflessione, Cconoscenze, Interventi, Arcadia Ricerche, Venice, Italy, 2005, pp. 487–496. http://www.arcadiaricerche.it/editoria/2005.htm.
[92] P. Antonioli, G. Zapparoli, P. Abbruscato, C. Sorlini, G. Ranalli, P.G. Righetti, Art-loving bugs: the resurrection of Spinello Aretino from Pisa's cemetery, Proteomics 5 (2005) 2453.
[93] F. Cappitelli, E. Zanardini, G. Ranalli, E. Mello, D. Daffonchio, C. Sorlini, Improved methodology for bioremoval of black crusts on historical stone artworks by use of sulfate-reducing bacteria, Appl. Environ. Microbiol. 72 (2006) 3733.
[94] F. Cappitelli, L. Toniolo, A. Sansonetti, D. Gulotta, G. Ranalli, E. Zanardini, C. Sorlini, Advantages of using microbial technology over traditional chemical technology in removal of black crusts from stone surfaces of historical monuments, Appl. Environ. Microbiol. 73 (2007) 5671.
[95] Bacteria that clean art: restorers and microbiologists use bacteria to make works of art shine like new, Asociación RUVID, ScienceDaily (June 7, 2011). http://www.sciencedaily.com/releases/2011/06/110607063411.htm.
[96] F. Valentini, A. Diamanti, M. Carbone, E.M. Bauer, G. Palleschi, New cleaning strategies based on carbon nanomaterials applied to the deteriorated marble surfaces: a comparative study with enzyme based treatments, Appl. Surf. Sci. 258 (2012) 5965.
[97] G. Lustrato, G. Alfano, A. Andreotti, M.P. Colombini, G. Ranalli, Fast biocleaning of medieval frescoes using viable bacterial cells, Int. Biodeterior. Biodegradation 69 (2012) 51.
[98] D. Kumar, Savitri, N. Thakur, R. Verma, T.C. Bhalla, Microbial proteases and applications as laundry detergent additives, Res. J. Microbiol. 3 (2008) 661.

Chapter 5

Cleanliness Verification on Large Surfaces

Instilling Confidence in Contact Angle Techniques

Darren L. Williams and Trisha M. O'Bryon
Sam Houston State University, Department of Chemistry, Huntsville, TX, USA

Chapter Outline

1. Background — 163
   1.1. Scope — 163
   1.2. Surface Cleanliness and Surface Energy — 164
2. Description of the Method — 167
   2.1. Traditional and Newly Digitized Contact Angle Methods — 167
       2.1.1. Side-on Methods (Half-angle, Drop Shape Analysis, and Snake) — 168
       2.1.2. A Newly Digitized Top–Down Method (Bikerman) — 170
       2.1.3. Reflected-Angle Methods (Langmuir) — 172
   2.2. Suggested SRO in the Literature or in Industry — 173
   2.3. SRO Materials — 173
3. Advantages and Disadvantages — 175
   3.1. Personnel Training — 175
   3.2. Method Comparisons — 175
4. Results — 175
   4.1. Examples of Imaging Choices — 175
   4.2. Example of Performance Comparison — 176
   4.3. Example of Personnel Training — 177
   4.4. Examples of Method Comparisons — 178
5. Applications — 179
6. Future Developments — 179
   Acknowledgments — 180
   References — 180

1. BACKGROUND

1.1. Scope

Cleanliness verification is growing in its importance in many industries, e.g. aerospace, biomedical engineering, and semiconductor fabrication [1], and the sessile drop technique stands out as an inexpensive, versatile, and portable way to probe the wettability of a surface [2], which is correlated with the cleanliness of hydrophilic (i.e. metallic and metallic oxide) surfaces. This chapter addresses the underserved community of users who clean surfaces that are too large to be placed in a small instrument or for whatever reason are not amenable to offline study. Hydrophobic oils, greases, and soils are their most common contaminants. This chapter may be less relevant to the user who is able to analyze small parts, wafers, and coupons using a commercial goniometer with subdegree accuracy and precision.

The ultimate goal is to facilitate the confident use of the sessile drop technique on large surfaces for contamination control and cleanliness verification. By "confident use," we mean techniques and procedures that possess verifiable accuracy and precision suitable for incorporation into a company's quality management infrastructure. By "cleanliness verification," we mean verification that the surface is suitably prepared for the next production step or end use. This is not necessarily the same as a contamination-free or pristine surface. Rather, knowledge of the surface energy required for optimum performance will be useful in writing specifications for cleanliness, and this chapter should assist in the verification that the desired surface energy specification has been achieved.

This chapter examines the utility of sessile drop contact angle measurement for surface energy determination and cleanliness verification. A review is given on the available methods, commercial instruments, patents, and literature describing the state of the art in contact angle measurement. Then, a description is given on contact angle measurement techniques that have been modified for use on large surfaces. The negative effects of these changes on accuracy and precision are discussed, and remedies are given including the use of standard reference objects (SROs) [3,4] that mimic the size and shape of sessile drops.

1.2. Surface Cleanliness and Surface Energy

Surface energy (free energy per unit area) and surface tension (force per unit length) are essential concepts for describing the characteristics of solid–liquid interactions [5]. A clean metal or metal oxide surface will typically have a high surface energy. Liquids, adhesives, and polymer melts will spontaneously coat a high-energy surface as long as the surface tension of the liquid is lower than the surface energy of the solid. Contamination, especially hydrocarbon soils, will lower the surface energy of the substrate leading to incomplete coating, adhesion failure, delamination, etc. Therefore, knowing the surface energy, types of soils, and surface tension characteristics of coatings and adhesives is essential for confident and functional cleanliness verification.

The surface tension of a liquid is typically measured directly using tensiometry, while the surface energy of a solid is determined indirectly using the wetting behavior of test liquids with known surface tensions [6]. The wetting behavior is easily measured using the contact angle θ of a sessile drop. The interplay of surface energy, surface tension, and contact angle has been described in great detail since the 1960s by Zisman [7] and many others.

Zisman developed a standard technique to determine the surface energy of a smooth, planar surface by studying the contact angle behavior of probe liquids of varying surface tension [7]. Although examples are given in Zisman's work of polar liquids and polar surfaces, the Zisman Plot uses a one-parameter approach to the surface tension of the liquid phase and surface energy of the solid substrate. Owens and Wendt [8], Rabel [9], and Kaelble [10] added a polar parameter, and eventually a hydrogen-bonding parameter to form a system similar to that of Hansen [11]. Alternatively, a description can be made in terms of dispersion and Lewis acid–base interactions as developed by van Oss, Chaudhury, and Good [12]. To save time in reviewing the above developments, the interested reader will find the review of these theories concisely delivered on the Krüss web site [13].

But for those new to surface energy determinations, an example of a simple Zisman analysis is given. In a Zisman plot, the cosine of the contact angle, cos θ, of each liquid is plotted against the surface tension of each liquid (γ_{lv}—the liquid–vapor interfacial tension). A line is fitted to the contact angle measurements and extrapolated to find the critical surface tension (γ_c) where spontaneous wetting occurs (i.e. cos θ = 1). Any liquid with a surface tension less than γ_c will completely wet the surface. Figure 5.1 is a Zisman plot generated using aqueous solutions of sodium dodecyl sulfate on an aluminum surface.

Although the Zisman plot is simple enough to analyze, we have found that a slight modification of the plot speeds the analysis considerably. In the modified Zisman plot (Fig. 5.2), the surface tension of each probe liquid is plotted versus

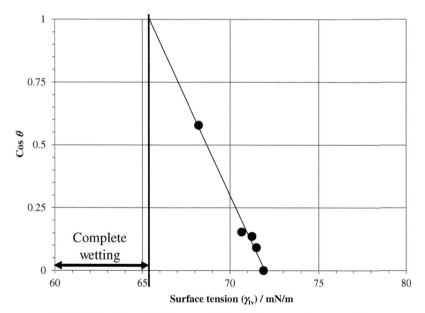

FIGURE 5.1 Zisman plot of aqueous SDS solutions on an aluminum surface.

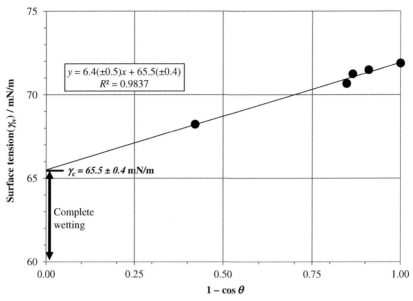

FIGURE 5.2 A modified Zisman plot of γ_{lv} versus $1 - \cos\theta$ allows the critical surface tension to be calculated directly from the y-intercept of a least squares fitting line.

"$1 - \cos\theta$". Almost all plotting packages are capable of displaying a trend line with a linear or polynomial fitting equation. The critical surface tension is the y-intercept in this modified plot. An analysis of variance routine on γ_{lv} vs $1 - \cos\theta$ yields the critical surface tension (65.5 mN/m) with an estimate of the standard error (0.4 mN/m). The modified plotting technique's ability to estimate the uncertainty in the critical surface tension is a distinct advantage over the traditional plotting technique.

If one is using very pure liquids, then the literature values of surface tension may be used. But if solutions are used as probe liquids, then one must measure the surface tension. The DuNouy tensiometer [14,15] uses a platinum–iridium ring that is pulled out of the vapor–liquid interface, and the force pulling on the ring is used to calculate γ_{lv}.

The authors have shown how a digital version of this tensiometer may be constructed from an analytical balance and a hydraulic press [16]. The balance must have a hook for weighing objects below the balance and must be able to communicate with a computer. The top platen of a Carver-type press typically has a hole in it. A ball chain may be allowed to hang through this hole from an analytical balance that is resting on the top platen. The other end of the ball chain holds the platinum–iridium ring. The lower platen holding the test solution is raised until the ring is submerged. Then, the hydraulic fluid in the press is released to slowly lower the liquid while the balance is recording the force on the ring as it passes through the liquid–air interface.

Chapter | 5 Cleanliness Verification on Large Surfaces

FIGURE 5.3 A sessile drop of 10 µL deionized water on a hydrophobic surface.

The ring tensiometry technique is easily checked against pure liquids to ensure that the instrumentation and the technician are producing accurate results. The weakness of all the surface energy analyses of Zisman et al. for cleanliness specification and verification lies in the uncertainty of the contact angle measurement. The contact angle is affected by surface contamination, roughness changes, surface tilt, liquid purity, liquid viscosity, surface reactivity, etc. Kumar and Prabhu review many of these factors in detail [17]. This strong dependence on the state of the surface illustrates the excellent sensitivity displayed by the sessile drop contact angle.

2. DESCRIPTION OF THE METHOD

2.1. Traditional and Newly Digitized Contact Angle Methods

Before discussing contact angle measurement validation, it is appropriate to describe some of the contact angle measurement techniques and analysis methods. By far the most common commercially available instruments view the drop profile with back illumination (Fig. 5.3). The analysis has been automated using computerized image analysis algorithms. These instrument vendors provide their own method validation procedures, and some supply validation slides (SROs) [3,4]. One drawback, however, is the inability of most of these commercial instruments to travel outside of the laboratory to the production floor, paint bay, or field.

Since most of the contact angle analysis methods are based on the geometry of a perfect sphere, one must use small drops on a level planar surface, although nonlevel and curved surfaces [18,19] have been addressed. According to Extrand and Moon [20] based on Eqn (5.1), a 10 µL water droplet will be spherical if it adopts a shape with a contact angle between 10° and 140°.

Equation (5.1) describes the maximum spherical volume (V_{max} in μL) in general terms suitable for any liquid where g is the acceleration due to gravity (9.81 m/s²), γ is the liquid surface tension in mN/m, and ϱ is the liquid density in g/cm³.

$$V_{max} = \frac{\pi}{48}\left(\frac{\gamma}{\rho g}\right)^{3/2} \tan(\theta/2)\left(3 + (\tan(\theta/2))^2\right)\left[\left(1 + 8\frac{(\sin\theta)^2}{1-\cos\theta}\right)^{1/2} - 1\right]^3$$

(5.1)

For models that depend on drop volume (e.g. Bikerman [21]), the uncertainty in drop volume is a large source of uncertainty in the resulting contact angle measurement. Some models (half-angle and Brugnara [22]) are insensitive to volume as long as the drop is spherical. Some drop shape analysis (DSA) routines (LB-ADSA [23]) model the gravity-induced shape, and still others (DropSnake [24]) do not depend on the shape of the drop at all. Each of these methods and their freely available software packages are described below.

2.1.1. Side-on Methods (Half-angle, Drop Shape Analysis, and Snake)

Measurement. For the most accurate view of the three-phase point, the camera should be placed perpendicular to the side of the drop (Fig. 5.4), which is elevated on a pedestal to bring it near the optical axis. In general, these goniometers consist of a light source (A), a mask, screen, or collimator (B), a high-resolution digital camera (C), and the sessile drop on a pedestal (D).

The use of a pedestal is not possible if the surface is very large. This side-on method has been modified using portable light sources and USB microscopes [25]. Figure 5.5 is a schematic of the side-on method using a digital microscope on a large surface. The light source (A) is behind a screen, mask, or collimator

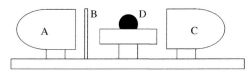

FIGURE 5.4 The apparatus for side-on contact angle measurement in most commercial instruments consists of a light source (A), a screen, mask, or collimator (B), an imaging device (C), and the sessile drop (D) on a pedestal that is elevated to the optical axis of the camera.

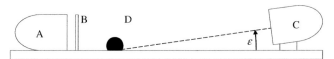

FIGURE 5.5 The apparatus for side-on contact angle measurement on a large surface consists of a light source (A), a screen, mask, or collimator (B), the sessile drop (D), and the imaging device (C) that is elevated above the surface a minimal amount.

(B) that reduces the amount of light reflecting off the apex of the drop into the camera (C). The angle of elevation (ε) should be small. However, small elevations require the camera distance to increase, which has the detrimental effect of reducing the pixel density across the drop image. The camera distance from the drop can be reduced through the use of a prism or mirror [25–27] (Fig. 5.6).

Analysis. Of all the image analysis methods, the half-angle method requires the least amount of effort. The contact angle of a drop, θ, on a surface is determined by Eqn (5.2) using the base, b, and the height of the drop, h, as seen in Fig. 5.7.

$$\theta = 2 \tan^{-1} \left(\frac{2h}{b} \right) \quad (5.2)$$

Almost any image processing software can be used to measure h and b, but the authors prefer to use ImageJ [28]—a freely available image processing platform that also contains plug-ins for contact angle determination [22–24].

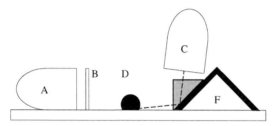

FIGURE 5.6 The apparatus for side-on contact angle measurement on a large surface using a mirror or prism (F), a light source (A), a screen, mask, or collimator (B), the sessile drop (D), and the imaging device (C).

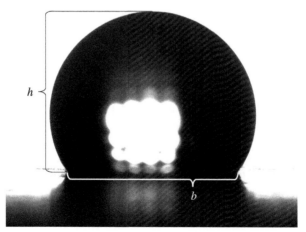

FIGURE 5.7 The measured parameters (h, b) needed for the half-angle contact angle determination method.

The major sources of uncertainty in the half-angle method arise from an inability to accurately identify the three-phase points and the apex of the drop. Often, one is unable to view the drop directly from the side, and is forced to look at the drop from a slightly elevated position above the surface (Figs 5.5 and 5.6). This introduces two biases into the analysis. For nonwetting drops, the base width (b) will appear to approach the drop diameter (d) as the camera is elevated. The height will appear to decrease sometimes, as the light reflecting off the top of the drop makes the identification of the drop apex uncertain. The net result of camera elevation is a contact angle biased toward 90°. This bias is mitigated by keeping the viewing elevation to a minimum and by using collimated light or a screen to prevent light from shining down onto the top of the drop. Likewise, for wetting drops, the contact angle will be biased toward 90° because the "sharpness" of the three-phase point is lost as the camera is elevated.

Despite its ease of use, the half-angle method described above depends on only three points to define the shape of the drop and the contact angle. There are more sophisticated methods for side-on image analysis that are freely available. These suffer from the same biases as the half-angle method with respect to view elevation, but their strengths are based on their ability to utilize more points along the drop edge. Their use and evaluation have been reported [25].

2.1.2. A Newly Digitized Top–Down Method (Bikerman)

Measurement. Bikerman postulated the idea of computing the contact angle of a sessile drop by measuring the diameter of the drop from above. This method for cleanliness testing was found to be useful on airplane fuselages— a perfect example of a surface that does not fit into a commercial goniometer [29–31].

He used a microscope fitted with a micrometer eyepiece to measure his drop diameters, but one can replace the traditional microscope with a digital microscope [32]. The digital microscope does not have a calibrated magnification, and thus, requires a calibration object to be placed in the field of view near the sessile drop. Figure 5.8 is a schematic of the top–down apparatus containing the illumination beam (A) from the digital microscope (C), the sessile drop (D), and the calibration object—a metal washer (G).

The calibration object may be dimensioned using a caliper. The image may be analyzed using any image measurement software, but some are ideal for measuring the diameter of circular objects in an image (e.g. Meazure [33]). A spreadsheet is useful for calibrating the image and computing the contact angle. A typical data image for top–down contact angle analysis is shown in Fig. 5.9.

Analysis. Bikerman derived the relationship between the contact angle θ, the base contact diameter b, and the volume v for a spherical drop (Eqn (5.3)). The base contact diameter is not visible from above if the contact angle is >90°. When the contact angle is >90°, Eqn (5.4) must be used because only the diameter of the drop d is visible from above [26]. Once b or d is measured, the ratio

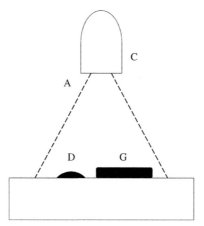

FIGURE 5.8 The top–down apparatus consists of an illumination beam (A) from the digital microscope (C), the sessile drop (D), and the calibration object (G).

FIGURE 5.9 An example of top–down image showing the calibration object (washer), a 10 µL drop of deionized water (top), and a 10 µL drop of 104 ppm SDS in water.

with v is computed and the contact angle, θ, is found numerically using a lookup table of Eqn (5.3) (or Eqn (5.4)) in a spreadsheet.

$$\frac{b^3}{v} = \frac{24 \sin^3 \theta}{\pi \left(2 - 3\cos\theta + \cos^3\theta\right)} \tag{5.3}$$

$$\frac{d^3}{v} = \frac{24}{\pi \left(2 - 3\cos\theta + \cos^3\theta\right)} \tag{5.4}$$

The top–down method decreases the size of the contact angle measurement apparatus, but requires an investment in a microsyringe or micropipette for accurate volume dispensing. Numerical analysis of Eqns (5.3) and (5.4) was used to determine that a 1% uncertainty in a 10 μL drop volume yields approximately a 1% uncertainty in contact angle in the spherically valid range of 10°–140°. A 1% uncertainty in drop diameter yields a slightly larger 1.5% uncertainty in contact angle over the same range. This method could be validated with a circular calibration object and a validated micropipette.

2.1.3. Reflected-Angle Methods (Langmuir)

Measurement and Analysis. The final method described here is the reflected-angle method first described by Irving Langmuir in 1937 [34]. Reflected-angle techniques have been used by a few instrument manufacturers as alternatives to the traditional side-on techniques. The Contact-θ-Meter [35] very closely matches the technique of Langmuir and is limited to a practical range of 10°–80°. A more recent and portable device (TVA100) uses reflected-angle analysis to measure the radius of curvature of the sessile drop. This radius and the drop volume may be used to calculate the contact angle in the range of 3.5°–75° [36].

The strength of the reflected-angle method is the direct computation of the contact angle ($\theta = 90° - 0.5\phi$) by the measurement of the reflected angle ϕ of a small beam of light from a fiber optic source (A) shining very close and parallel to the surface (Fig. 5.10).

The downside of the Contact-θ-Meter is the same as the lab-based goniometers, namely, they only accept small coupons as sample surfaces. However, we have successfully used cell phone cameras for the Bikerman method [29] and, with the internal accelerometers of "smart phones," one can measure the angle of the reflected beam (*R*) in Fig. 5.10 [26].

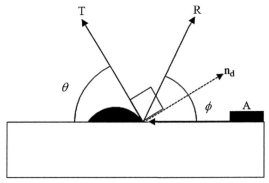

FIGURE 5.10 The reflected-angle technique utilizes the angle ϕ of the illumination source's (A) reflection (*R*) off the drop near the three-phase point. This reflection is symmetric about the normal from the drop surface (n_d), which is orthogonal to the contact-angle-defining tangent (*T*).

Validation of Langmuir's reflected-angle methods presents a problem. The two-dimensional standard reference objects (2D SROs) are not suitable because they are merely profile images of sessile drops. A high-fidelity three-dimensional sessile drop standard reference object—a 3D SRO—is needed so that the reflective geometry can be used to validate method and technician performance.

2.2. Suggested SRO in the Literature or in Industry

The concept for a >90° 3D contact angle standard has been mentioned, for instance in ASTM D 5725-99 [37]. But this and other mentions in the literature did not contain enough detail to instill confidence in the production and use of such an SRO. Some instrument manufacturers provide [3,4] high-quality contact angle images imprinted on glass slides which are placed on the goniometer stage in place of an actual sessile drop (Fig. 5.11). These slides are examples of a 2D SRO, and they work well in situations where the illumination, sample pedestal, and camera have a stable, side-on geometry. But these slides are less useful if one is adapting the illumination and camera positions to enhance the contrast and focus of an actual sessile drop on a large surface that cannot be brought to the lab. These slides cannot be used for top–down or reflected-angle methods, either. In general, it is the best when the calibration object closely mimics the size, shape, and reflective characteristics of an actual sessile drop.

2.3. SRO Materials

A >90° contact angle standard can be constructed using a 3.18 mm chromium steel ball (MSC Industrial Supply Co. #72660) mounted in various drill gage card

FIGURE 5.11 The 2D standard reference object from Ramé-Hart showing four calibrated sessile drop profile images of 31.5° (A), 61.0° (B), 90.0° (C), and 119.5° (D).

(Grainger #5C732) holes to mimic a series of contact angles (Fig. 5.12). Less than 90° contact angle standards can be constructed using 6.35 mm and 9.53 mm diameter balls (MSC Industrial Supply Co. #72702 and #72744) mounted under small sheets of punched aluminum (Fig. 5.13). The ball diameters and holes were chosen to produce faux sessile drops with volumes near 15 µL.

FIGURE 5.12 A nominal 119° SRO constructed using a 3.18 mm chromium steel ball in a 2.77 mm drilled gauge hole viewed from above and from the side using a prism.

FIGURE 5.13 A nominal 54.3° SRO constructed using a 6.35 mm chromium steel ball in a 5.16 mm hole drilled in galvanized metal sheeting viewed from above and from the side using a prism.

These types of standards have proven to be useful in comparing four different side-on methods and multiple student technicians [25]. The main advantages of these standards are the fact that they are rigid, nonevaporating, very spherical in shape, portable, and resistant to damage.

The <90° SROs exhibit one drawback. Since the ball is protruding through a cylindrical hole, there is a gap near the three-phase point. A machined hole with a matching spherical contour along the walls of the hole would eliminate this gap almost completely.

The possibility of printing a 3D SRO was explored using a uPrintSE (Stratasys, Inc) [38]. It was found that the extrusion nozzle of our 3D printer was too large for a high-fidelity reproduction of a 10 µL sessile drop. It was apparent to the naked eye, and under microscopic inspection, the drops were very rough and resembled bee hives.

3. ADVANTAGES AND DISADVANTAGES

3.1. Personnel Training

A select set of sessile drop images can be used to test the operator's competence with the software. This is a technique used at our university to ensure that new research students are able to precisely and accurately analyze the sessile drop image data. But there is more to contact angle analysis than image analysis.

A realistic 3D SRO allows the evaluation of an operator's ability to align, illuminate, and capture high-quality images of sessile drops. We have found that drop illumination and camera elevation angle are the two skills with the steepest learning curve. The SRO provides an objective target to this seemingly subjective category of image quality. A quality image produces an accurate contact angle, and accurate contact angles instill confidence in surface cleanliness decisions.

3.2. Method Comparisons

Strength of the 3D SRO is the ability to compare methods. The ball in a hole allows the comparison of all the side-on methods. But this SRO is not appropriate for the top–down and reflected-angle methods. The ball protruding through a hole allows the comparison of all the methods described in this chapter.

4. RESULTS

4.1. Examples of Imaging Choices

The choice between a glass prism (Edmund Optics) and a bent electropolished metal mirror (Rimex Inc, Edison, NJ) was evaluated against the Ramé-Hart 2D SRO. The metal mirror is preferred for cost and durability factors. As an example, the results are shown for the 119.5° drop image in Table 5.1. It is

TABLE 5.1 Comparison of the Performance of the Mirror vs the Prism

	Accuracy: Average Error/°	Precision: Standard Deviation/°	N
Mirror	−2.2	0.6	6
Prism	0.38	0.40	6

N: number of samples.

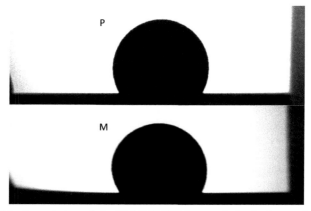

FIGURE 5.14 A comparison of the 2D 119.5° SRO images obtained using a prism (P) and a bent-metal mirror (M).

important to note that the uncertainty with the mirror is only slightly more than the prism. However, the accuracy is compromised by the slightly distorted image of the bent metal mirror (Fig. 5.14). This judgment is impossible unless one can use an SRO to calculate the accuracy values, and therein lies the whole motivation for using an SRO in a cleanliness verification quality management plan.

4.2. Example of Performance Comparison

The prism was used in the arrangement shown in Fig. 5.6 where the sessile drop was replaced by the 2D SRO (Fig. 5.11). Three images of each drop profile were analyzed in ImageJ using the half-angle method. The absolute error was calculated by subtracting the accepted value from the experimental value. Table 5.2 shows the accuracy of this imaging method in terms of the mean absolute error and the precision of this imaging method in terms of the standard deviation of each set of three observations. The pooled standard deviation for this imaging method is 0.3°.

Chapter | 5 Cleanliness Verification on Large Surfaces

TABLE 5.2 Performance of the Prism Imaging Method against the 2D SRO

θ/°	Accuracy: Mean Error/°	Precision: Standard Deviation/°	N
31.5	−2.3	0.5	3
61.0	−2.0	0.2	3
90.0	−1.2	0.2	3
119.5	0.1	0.2	3

N: number of samples.

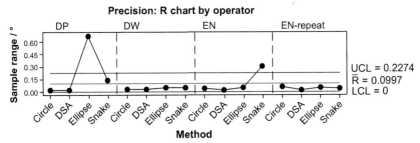

FIGURE 5.15 The use of a 3D SRO to evaluate the results obtained by four operators (DP, DW, EN, and EN-repeat) using four contact angle measurement procedures (Circle, DSA, Ellipse, and Snake). The horizontal reference lines in each chart are the 95% upper confidence limit (UCL), the mean of all measurements in the study (X), the 95% lower confidence limit (LCL), and the mean of all ranges in the study (R). A color version of this figure appears in the color plate section.

4.3. Example of Personnel Training

The 3D SROs were used to evaluate the performance of three operators (DP, DW, and EN). The operators varied in their experience from over 1 year (DW) to 1 week (EN) to 1 day (DP). The contact angle measurement methods were the Circle and Ellipse methods of Brugnara and the DSA and Snake methods of Sage [25]. The accepted value for the SRO was determined using the calibrated half-angle method. Figure 5.15 shows the accuracy and the precision of the operators and methods.

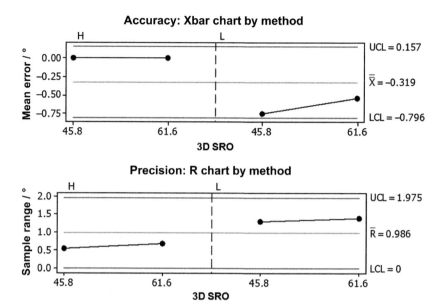

FIGURE 5.16 The half-angle method (H) was used to determine the accuracy and precision of the Langmuir method (L) against two 3D SROs (45.8° and 61.6°). The horizontal reference lines in each chart are the upper 95% confidence limit (UCL), the mean of all measurements in the study (X), the lower 95% confidence limit (LCL), and the mean of all ranges in the study (R). A color version of this figure appears in the color plate section.

Clearly, DP had accuracy and precision problems with the Ellipse method and should be retrained. The operator NE initially had difficulty with the Snake method, but self-corrected when the analyses were repeated. The ability to evaluate accuracy objectively using an SRO instills confidence that new operators are performing within acceptable limits.

4.4. Examples of Method Comparisons

The Langmuir method described herein has been evaluated against two 3D SROs constructed to exhibit contact angles of 48.1° and 61.0°. The half-angle analysis of these SROs gives the accepted values of 45.8 ± 0.2° and 61.6 ± 0.3° ($N = 6$ each). Figure 5.16 shows the comparison of the accuracy and precision of the Langmuir (L) method to the half-angle (H) method against these standards.

The Langmuir (L) method matched the half-angle (H) method to within a degree (Fig. 5.16). The mean error of this method was slightly below zero at −0.6 ± 0.5° ($N = 12$). The standard deviation of this method (0.5°) is acceptable for many operations, and is comparable to the standard deviation of the half-angle method ($s = 0.2°$).

5. APPLICATIONS

The uncertainty in cleanliness verification is tied to the uncertainties in surface energy determination. There are several methods for determining surface energy, and all of them use test liquids of known (or measured) surface tensions along with the contact angles these liquids make with the surface. These contact angle measurements are the largest source of uncertainty in the surface energy determination. Additionally, measurements of contact angles on large surfaces present even more challenges. We have endeavored to review three newly modified techniques suitable for large surface contact angle measurement. And we have presented, demonstrated, and evaluated several options for producing 3D SROs that are suitable for use in industrial cleanliness verification activities.

The use of steel plates, gauge cards, and bearing balls as robust standards shows promise. The ball resting in or protruding through a hole is the best approach. We have found that drilled holes exhibit rough edges and slight non-circularity. Punched holes are preferred for the >90° 3D SROs, although these punched holes show indentation near the edge of the hole. This defect seems more manageable in side-on techniques. Machined holes are necessary for the protruding ball (<90°) SROs.

These SROs are suitable for validating the illumination, sample, and imaging setup in field applications or on the large-part manufacturing floor where true-profile, side-on imaging is impossible. An SRO also gives the user the ability to check accuracy and not merely precision, which has been a long-time difficulty of contact angle method comparisons. Performance can also be monitored across multiple facilities and across long periods of time with proceduralized SRO checks.

The combination of these validation tools and the modified contact angle measuring techniques fills a need for robust, production-line capable of cleanliness verification methods.

6. FUTURE DEVELOPMENTS

The connection between cleanliness and surface energy is well established. Ironically, the strength and weakness in this analysis is the sensitivity of the sessile drop contact angle. A 3D SRO that mimics a sessile drop is an effective quality assurance tool for eliminating non-process-related variation in the contact angle results.

Admittedly, the 3D SROs shown here are at a very primitive level of development. Now that the proof of concept is complete, high-precision machined base plates with permanently mounted bearing balls may be produced. It is likely that certified 3D SROs may become available in the near future.

ACKNOWLEDGMENTS

The Welch Foundation Departmental Development Grant is acknowledged for funding much of this work. Many students who have been supported by this grant to do contact angle work are acknowledged for their efforts throughout the years. They are Mark Amann, Jennifer Bradley, Madison Hausinger, James Huskey, Megan Konarik, Elizabeth Nesselrode, Dustin Palm, and Angela Rippley. Our colleague Anselm Kuhn is acknowledged for supplying the prisms and the bent-metal mirrors from Rimex Metals.

REFERENCES

[1] B. Kanegsberg, E. Kanegsberg, Handbook for Critical Cleaning, second ed., Applications, Processes, and Controls. vol. II, CRC Press, Taylor and Francis Group, Boca Raton, FL, 2011.
[2] W. Birch, A. Carré, K.L. Mittal, Wettability techniques to monitor the cleanliness of surfaces, in: R. Kohli, K.L. Mittal (Eds.), Developments in Surface Contamination and Cleaning, William Andrew Publishing, Norwich, NY, 2008, pp. 693–723.
[3] Krüss GmbH, Hamburg, Germany, List of Accessories (Contact Angle Measuring Instruments) (accessed 11.12.12). http://www.kruss.de/en/products/contact-angle/accessories/dsa100.html.
[4] Ramé-Hart Instrument Co, Succasunna, NJ, Calibration Tools (accessed 11.12.12). http://ramehart.com/calibration_tools.htm.
[5] A.W. Adamson, A.P. Gast, Physical Chemistry of Surfaces, sixth ed., Wiley Interscience, New York, NY, 1997.
[6] F.M. Etzler, Characterization of surface free energies and surface chemistry of solids, in: K.L. Mittal (Ed.), Contact Angle, Wettability and Adhesion, vol. 3, VSP, Utrecht, The Netherlands, 2003, pp. 219–264.
[7] W.A. Zisman, Relation of the equilibrium contact angle to liquid and solid constitution, Contact Angle, Wettability, and Adhesion, American Chemical Society, Washington, D.C, 1964, pp. 1–51. Advances in Chemistry Series No. 43.
[8] D.K. Owens, R.C. Wendt, Estimation of the surface free energy of polymers, J. Appl. Polym. Sci. 13 (1969) 1741.
[9] W. Rabel, Wetting theory and its application to the study and use of the surface properties of polymers, Farbe + Lack 77 (1972) 997.
[10] D.H. Kaelble, Dispersion-polar surface tension properties of organic solid, J. Adhes. 2 (1970) 66.
[11] C.M. Hansen, Hansen Solubility Parameters: A User's Handbook, second ed., CRC Press, Boca Raton, FL, 2006.
[12] C.J. van Oss, M.K. Chaudhury, R.J. Good, Interfacial Lifshitz-van der Waals and polar interactions in macroscopic systems, Chem. Rev. 88 (1988) 927.
[13] Krüss GmbH, Hamburg, Germany, Contact Angle and Surface Energy (accessed 11.12.12). http://www.kruss.de/en/theory/measurements/contact-angle.html.
[14] W.D. Harkins, H.F. Jordan, A method for the determination of surface and interfacial tension from the maximum pull on a ring, J. Am. Chem. Soc. 52 (1930) 1751.
[15] C. Huh, S.G. Mason, A rigorous theory of ring tensiometry, Colloid Polym. Sci. 253 (1975) 566.
[16] D.L. Williams, C.L. Jupe, K.D. Kuklenz, T.J. Flaherty, An inexpensive, digital instrument for surface tension, interfacial tension, and density determination, Ind. Eng. Chem. Res. 47 (2008) 4286.

[17] G. Kumar, K.N. Prabhu, Review of non-reactive and reactive wetting of liquids on surfaces, Adv. Colloid Interface Sci. 133 (2007) 61.
[18] M. Blitshteyn, J. Hansen, R. K. Shaw, Method and apparatus for determining the contact angle of liquid droplets on curved substrate surfaces, U.S. Patent 5,137,352, 1991.
[19] M. Guilizzoni, Drop shape visualization and contact angle measurement on curved surfaces, J. Colloid Interface Sci. 364 (2011) 230.
[20] C.W. Extrand, S.I. Moon, When sessile drops are no longer small: transitions from spherical to fully flattened, Langmuir 26 (2010) 11815.
[21] J.J. Bikerman, A method of measuring contact angles, Ind. Eng. Chem. 13 (1941) 443.
[22] M. Brugnara, Contact Angle Plugin for ImageJ Software, University of Trento, Trento, Italy, 2010.
[23] A.F. Stalder, T. Mechior, M. Müller, D. Sage, T. Blu, M. Unser, Low-bond axisymmetric drop shape analysis for surface tension and contact angle measurements of sessile drops, Colloids Surf. A 364 (2010) 72.
[24] A.F. Stalder, G. Kulik, D. Sage, L. Barbieri, P. Hoffmann, A snake-based approach to accurate determination of both contact points and contact angles, Colloids Surf. A 286 (2006) 92.
[25] D.L. Williams, A.T. Kuhn, M.A. Amann, M.B. Hausinger, M.M. Konarik, E.I. Nesselrode, Computerised measurement of contact angle, Galvanotechnik 108 (2010) 2502.
[26] T.M. O'Bryon, Preparation of a Non-Evaporating Contact Angle Standard Reference Material, M.S. Thesis, Sam Houston State University, Huntsville, TX, 2012.
[27] E. Wallström, I. Svenningsen, Surface Tension Measurements of Corrosion Resistant Paints – Method Study, Report T10–81M Scandinavian Paint and Printing Ink Research Institute, Horsholm, Denmark, 1981.
[28] W.S. Rasband, ImageJ Software, U. S. National Institute of Mental Health, Bethesda, MD, 2010.
[29] R.N. Miller, Rapid method for determining the degree of cleanliness of metal surfaces, Mater. Protect. Perform. 12 (1973) 31.
[30] R.N. Miller, Method for measuring surface cleanliness, U.S. Patent 3,618,374, 1971.
[31] J.B. Durkee, A.T. Kuhn, Wettability measurements and cleanliness evaluation without Substantial cost, in: K.L. Mittal (Ed.), Contact Angle, Wettability and Adhesion, vol. 5, VSP/Brill, Leiden, The Netherlands, 2008, pp. 115–138.
[32] D.L. Williams, A.T. Kuhn, T.M. O'Bryon, M.M. Konarik, J.E. Huskey, Contact angle measurements using cell phone cameras to implement the Bikerman method, Galvanotechnik 109 (2011) 1718.
[33] B. Roberts, Meazure Ver. 2.0 Build 158 Software, C-Thing Software, Mountain View, CA, 2004.
[34] I. Langmuir, V.J. Schaefer, The effect of dissolved salts on insoluble monolayers, J. Am. Chem. Soc. 59 (1937) 2400.
[35] Livereel Contact-θ-meter, Pearson Panke Equipment Ltd., London, UK.
[36] Krüss GmbH, Hamburg, Germany, TVA100 Contact Angle Measuring Module (accessed 11.12.12). http://www.kruss.de/en/products/contact-angle/top-view-contact-angle-analyzer-module-tva100.html.
[37] ASTM D5725-99(2008), Standard Test Method for Surface Wettability and Absorbency of Sheeted Materials Using an Automated Contact Angle Tester, (Withdrawn 2010), American Society for Testing and Materials, Conshohocken, PA, 2008. Replaced by TAPPI T 558 Om-10, Surface Wettability Absorbency Sheeted Mater. Using Automated Contact Angle Tester, Tech. Assoc. Pulp Paper Ind., Norcross, GA, 2010.
[38] Specification Sheet uPrintSE, Stratasys Inc., Eden Prairie, MN (accessed 11.12.12) URL http://www.uprint3dprinting.com/pdfs/specs/uPrintSE_SEPlus_3DPrintPack.pdf.

Index

Note: Page numbers with "f" denote figures; "t" tables.

A

Adhesion forces, 108
Advanced spray development, 126–133
    damage threshold, 129–133, 129f–130f
    droplet energy density, 127–128, 128f–129f
    nozzle development, 126–127, 127f
    See also Dual-fluid spray cleaning technique
Aerobic microorganisms, 143
Aerosol spray cleaning, 116
Alkyl groups, 8
American Type Culture Collection (ATCC), 141
Ammonia peroxide mixtures (APM), 115, 122–123
Ammoniums, 6t
    quaternary, 7f
    toxicity of, 25f
Anaerobic microorganisms, 143
Anions, 6t, 7–8
Aprotic ionic liquids (AILs), 9–10
Archaea, 141
Art objects/structures, cleaning of, 155
Artworks, cleaning of
    using ionic liquids, 39
    using microemulsions, 97–98
Atomic force microscopy (AFM), 113–114, 114f, 130
Azoliums, 6t

B

Bacillus, 142
Bacteria, 141
Bacterial characterization, microbial cleaning for, 153
Batch-type wafer cleaning tools, 116
Biocatalyst, 143
Bioremediation, 140–141
Black salt crusts, cleaning of, 155
Borate, 6t
Brush cleaning, 33, 34f–35f
Building exteriors, microemulsion cleaning of, 97

C

Calcium sulfate deposits, removal of, 142
Carbon, as nutrients for microbes, 142–143
Carbonate, 6t
Carbonic acids, 6t
Cations, 5–7, 6t, 7f
Cellulose acetate-replicating tape, 153, 153f
CH completion wells, formation damage removal in, 92–93
Chemical particle removal methods, 115
Cleaning process window, 108–115, 110f, 115f
    experimental studies, 113–115
    forces acting on 100 nm particle in solution, 110t
    theoretical predictions, 110–113
Cleaning tank, before and after microbial cleaning, 152f
Cleanliness, surface
    levels, surface contamination and, 140
    monitoring of, 153
Compositions, microbial, 145
Contact angle, of microemulsions, 76–77, 76f–77f
Contact angle measurement, 164, 167
    traditional and newly digitized methods, 167–173
    newly digitized top–down method, 170–172
    reflected-angle methods, 172–173
    side-on methods, 168–170
Contact lenses, cleaning of, 155
Contact-θ-Meter, 172
Contaminant(s)
    removal, with microemulsions, 81
    types of, 145–146

Contaminated soil, microemulsion cleaning of, 94–96, 96f
Costs
    comparison of conventional solvent process with microbial cleaning, 149t
    of microbial cleaning, 148–149
    savings, 149
    of surface contaminants removal using ionic liquids, 29–30
Cosurfactant effect, on microemulsion formulation, 71
Critical particle diameter, 107–108, 109t
Crown formation *see* Liquids surface, impact of droplets on
Crude oil reservoirs, microemulsion cleaning of, 98–99
Cyanate, 6t

D

Damage threshold
    advanced spray development, 129–133, 129f–130f
    relationship between droplet energy density and PRE, 131, 131f
    relationship between removed particle counts versus droplet velocity, 130, 131f
Debridement, microbial cleaning for, 154
Deep eutectic solvents (DES), 26–27, 30, 32–33, 36–38, 41
    choline chloride-based, 26t, 27f
    comparison with ionic liquids, 28t
Desulfovibrio desulfuricans, 142
Desulfovibrio vulgaris, 142
Differential scanning calorimetry, 82
Disinfection, microbial cleaning for, 154–155
Droplet energy density
    advanced spray development, 127–128, 128f
    correlation with number of damage sites, 129f
    relationship with PRE, 131, 131f
Droplet impact energy, 117–121
    crown formation, 119–120, 119f
    liquid film, impact on, 120–121, 121f–122f
    solid surface, impact on, 118–119, 118f
Droplet jet technique, 111, 111f
Drop shape analysis (DSA), 168
    See also Side-on contact angle methods
Dual-fluid spray cleaning technique, 107–138
    advanced spray development, 126–133
    cleaning process window, 108–115

droplet impact energy, 117–121
dual-fluid spray development, 122–126
particle removal techniques, overview of, 115–116
particles and adhesion forces, 108
spray nozzles, 117, 117f
system description, 116–117
DuNouy tensiometer, 166

E

Electropolishing, 36, 36f
Energy, surface, 164–167
Environmental Protection Agency, 140–141
Equivalent alkane carbon number (EACN), 69, 73, 81

F

Fracturing gels from shale, microemulsion cleaning of, 99
Frescoes, microemulsion cleaning of, 97–98
Froth flotation, microemulsion, 94
Fuel desulfurization, ionic liquids applications in, 42
Fungi, 141

G

Gas wells, near-wellbore cleaning in, 88–93
    CH completion wells, formation damage removal in, 92–93
    OH completion wells, oil-based fluid filter cake removal in, 89–92
Glutaraldehyde, 155
Grease removal, microbial cleaning for, 151–152, 152f
Groundwater, microemulsion cleaning of, 94–96
Guanidiniums, 6t, 7f

H

Half-angle contact angle determination method, 169–170, 169f, 177–179, 178f
    See also Side-on contact angle methods
Halide, 6t
Hazardous materials decontamination, cleaning for, 37–38
Historical art objects/structures, cleaning of, 155
Household applications, cleaning for, 156

Index

Hydrocarbon(s)
  production, ionic liquids applications in, 40–41
  removal of, 142–143, 145
  soils, 164
Hydrochlorofluorocarbons (HCFCs), 2, 3t, 139–140
Hydrogen sulfide generation, 152–153
Hydrophilic–lipophilic deviation (HLD), 70

I

Imidazoliums, 6t, 7f
  toxicity of, 25f
Imide, 6t
Institutional applications, cleaning for, 156
Interfacial tension (IFT), 70–71, 79–85, 89, 93, 96–98, 100
  cosurfactants effect on, 71, 72f
  of microemulsion systems, 75, 75f
International Technology Roadmap for Semiconductors (ITRS), 107–108
  front end surface preparation roadmap for critical particle size and number, 109t
Ionic liquids (ILs), 4–27
  abbreviations and nomenclature of, 5–8
    alkyl groups, 8
    anions, 6t, 7–8
    cations, 5–7, 6t, 7f
  applications of, 4f
  aprotic, 9–10
  background of, 5
  characteristics of, 9–13, 10t
  comparison with deep eutectic solvents, 28t
  data compilations of, 24–26
  electrical conductance of, 20–22
  high-vacuum analytical applications of, 22–23
  metal-based, 12f
  physical appearance of, 11f
  protic, 9–11, 12f
  solubility of, 17–18
  surface contaminants removal using *see* Surface contaminants removal using ionic liquids
  thermal properties of, 14
  thermodynamic properties, modeling and predictions of, 18–19
  toxicity of, 23–24
  viscosity, 19–20
  volatility of, 16

L

Langmuir method *see* Reflected-angle contact angle methods
Large surfaces, cleanliness verification on, 163–182
  applications of, 179
  future developments of, 179
  imaging choices, 175–176, 176t
  method comparisons, 175, 178
  performance comparison, 176, 177t
  personnel training, 175, 177–178
  scope, 163–164
  standard reference objects, 173
    materials, 173–175
  surface energy, 164–167
  traditional and newly digitized contact angle methods, 167–173
  See also Sessile drop technique
Laundry detergents, 156
Linkers
  effect on microemulsion formulation, 71–73
  hydrophilic, 71–72
  lipophilic, 71–72
Liquid film, droplets impact on, 120–121, 121f–122f
Liquids surface, droplets impact on, 119–120, 119f–120f

M

Megasonics, 116
Mercury bioremediation, microbial cleaning for, 153–154
Metal halide, 6t
Methicillin-resistant *Staphylococcus aureus* (MRSA), 154
Methide, 6t
Microbes, application of, 145
Microbial agents, 141–142
Microbial cleaning, 139–162
  advantages and disadvantages of, 149–151
  applications of, 151–156
  cleanliness levels of, 140
  contaminants, types of, 145–146
  costs of, 148–149
  life cycle diagram of, 142f
  microbial agents of, 141–142
  principles of, 142–143
  substrates, types of, 146
  systems, 143–149
Microbial contamination, cleaning of, 38
Microbial mergers, 141

Microemulsions
    cleaning applications of, 65–106
        artwork, 97–98
        basic principles of, 77–80, 80f
        building exteriors, 97
        cleaners and evaluation techniques, design of, 81–82
        contaminant removal, 81
        contaminated soil, 94–96, 96f
        crude oil reservoirs, 98–99
        current trends of, 100
        fracturing gels from shale, 99
        frescoes, 97–98
        froth flotation, 94
        future developments of, 100
        groundwater, 94–96
        oil and gas wells, near-wellbore cleaning in, 88–93
        oil-contaminated drill cuttings, 84–85, 84f, 85t, 86f
        surface cleaning, 81
        textiles, 96–97
        using nonaqueous solvents, 97
        wastewater cleaning, 94, 95f
    formulations, 67–74
        cosurfactant effect, 71
        linkers effect, 71–73
        oil or solvent, type of, 73
        salinity effect, 70–71
        surfactant selection, 70
        temperature effect, 74
    properties of, 74–77
        contact angle, 76–77, 76f–77f
        interfacial tension, 75, 75f
        solubilization, 74
        wettability, 76–77
M-jet scrubber, 122
Montreal Protocol, 2, 3t

N

NanosprayÅ nozzle, 126–128, 130
    droplet size and velocity distributions of, 127f
    dual-fluid, threshold curves for particle removal and pattern damage generation with, 132–133, 132f
    images of droplets from, 127f
    PRE performance for removal of PSL spheres, 133f
Nanospray2 nozzle, 126
    damage threshold curve with droplets, 130, 130f
    droplet size and velocity distributions of, 127f
    dual-fluid, threshold curves for particle removal and pattern damage generation with, 132–133, 132f
    images of droplets from, 127f
    PRE performance for removal of PSL spheres, 133f
Nanospray nozzle, 124–126, 125f–126f
    droplet size distribution and nitrogen flow effect on droplet size for, 125f
    See also NanosprayÅ nozzle; Nanospray2 nozzle
Near-wellbore cleaning in oil and gas wells, 88–93
    CH completion wells, formation damage removal in, 92–93
    OH completion wells, oil-based fluid filter cake removal in, 89–92
Nonaqueous solvents, microemulsion cleaning using, 97
Nonvolatile residue (NVR), 140
Nozzle
    advanced spray development, 126–127, 127f
    See also specific entries
Nuclear magnetic resonance (NMR), 82

O

OH completion wells, oil-based fluid filter cake removal in, 89–92
Oil-based drilling fluids to water-based fluid displacement, wellbore cleanup during, 85–88
Oil-contaminated drill cuttings, cleaning of, 84–85, 84f, 85t, 86f
Oil-contaminated sands, cleaning of, 37
Oilfields, sulfate-reducing bacteria in, 152–153
Oil removal, microbial cleaning for, 151–152
Oil wells, near-wellbore cleaning in, 88–93
    CH completion wells, formation damage removal in, 92–93
    OH completion wells, oil-based fluid filter cake removal in, 89–92
Organic solvents, characteristics of, 10t
Other ozone-depleting solvents (ODCs), 2

P

Particle removal efficiency (PRE), 115–116, 124
    relationship with droplet energy density, 131, 131f

Index

results from removal of PSL spheres, 133–134, 133f–134f
for silicon nitride particles removal, 123f
Particles, 108
removal *see* Dual-fluid spray cleaning technique
Particulate matter, cleaning of, 37
Parts clean(ers/ing), 33–36, 143–148, 144f, 151
cleaning of parts in wash basin, 147f
operating guidelines, 147–148
sink before and after cleaning, 147f
Peracetic acid, 155
Phase Doppler Particle Analysis technique, 124
Phosphate, 6t
Phosphoniums, 6t, 7f
Physical particle removal methods, 115–116
Physicochemical cleaning effect, 122–123
Piperidiniums, 6t, 7f
Polystyrene latex (PSL) particles, 113–114, 122
spheres, PRE results from removal of, 133–134, 133f–134f
Precision cleanliness level, 140
Prism imaging method, 176f
versus 2D SRO, performance comparison between, 177t
Protein-enhanced surfactants, 152
Protic ionic liquids (PILs), 9–11, 12f, 33–35
Protista, 141
Pseudomonas, 142
Pseudomonas stutzeri, 155, 156f
Pyrazoliums, 6t
Pyridiniums, 6t, 7f
Pyrrolidiniums, 6t

Q

Quantitative structure–activity relationships (QSARs), 24

R

RCA Standard Clean, 115
Reflected-angle contact angle methods, 172–173, 172f, 178, 178f
Ring tensiometry technique, 167
Roll brush scrubbing, 122

S

Salinity effect, on microemulsion formulation, 70–71
Semiconductor cleaning, 32–33
Sessile drop technique, 164
See also Large surfaces, cleanliness verification on
Severe acute respiratory syndrome, 154
Shock waves, 111–112, 111f, 113f, 120f
Side-on contact angle methods, 168–170
apparatus for, 168f
using mirror or prism, 169f
Single-wafer processing tools, 116
Small-angle neutron scattering, 82
Snake method, 178
See also Side-on contact angle methods
Soft Spray nozzle, 122–123, 123f, 125, 125f
droplet size distribution and nitrogen flow effect on droplet size for, 125f
Solid surface, droplets impact on, 118–119, 118f
Solubilization of microemulsions, 74
Solutions, microbial cleaning, 145
Solvent-based cleaning
cost comparison with microbial cleaning, 149t
Solvent cleaning, 139–140
Spray cleaning nozzles, 117, 117f
Spray development
advanced spray, 126–133
dual-fluid, 122–126
Standard reference objects (SROs), 164, 173
materials, 173–175
nominal 54.3°, 174f
nominal 119°, 174f
for reflected-angle contact angle methods, 173
3D. *See* 3D standard reference objects
2D. *See* 2D standard reference objects
STAR-CD software, 130
Substrates, types of, 146
Sulfate, 6t
Sulfate-reducing bacteria (SRB), 142
in oilfields, 152–153
Sulfoniums, 6t, 7f
Supercritical gases, cleaning with, 36–37
Surface cleanliness
levels of, 2–4
with microemulsions, 81
monitoring, microbial cleaning for, 153
Surface contaminants removal using ionic liquids, 1–64
advantages of, 30–31
applications of, 32–42
artworks, 39
brush cleaning, 33, 34f–35f
cleaning in place, 38

Surface contaminants removal using ionic liquids (*Continued*)
  consumer product applications, 42
  electropolishing, 36, 36f
  fuel desulfurization, 42
  hazardous materials, decontamination of, 37–38
  hydrocarbon production, 40–41
  microbial contamination, 38
  oil-contaminated sands, 37
  particulate matter, 37
  parts cleaning, 33–36
  semiconductor cleaning, 32–33
  supercritical gases, 36–37
 basic principles, 28
 cost of, 29–30
 disadvantages of, 31–32
Surface energy, 164–167
Surface tension, 164
Surfactant(s)
 extended, 73
 selection, for microemulsion formulation, 70
Surfactant affinity difference (SAD), 69–70
Surgical instruments, cleaning of, 155

T

Temperature effect, on microemulsion formulation, 74
Tensiometry, 164, 166
 digital version of, 166
 ring tensiometry technique, 167
Textiles, microemulsion cleaning of, 96–97
Thiazolium, 7f
Thiouroniums, 6t
3D standard reference objects, 175, 177, 177f, 179
Top–down contact angle method, digitized, 170–172, 171f
Tosylate, 7f
Toxic Substances Control Act (TOSCA), 141
Triazolium, 7f
Trichloroethane, 2
Triflate, 7f
Truck fueling bay, before and after microbial cleaning, 152f
TVA100, 172

2D standard reference objects, 173, 173f, 175–176
 images comparison using prism and mirror, 176f
 vs. prism imaging method, performance comparison between, 177t

U

Ultracentrifugation, 82
Ultrasonics technique, 111, 111f
uPrintSE, 175
Uroniums, 6t
U.S. Department of Health, Public Health Service, 141

V

Viruses, 141

W

Wastewater, microemulsion cleaning of, 94, 95f
Weber number, 117
Wellbore cleaning
 near-wellbore cleaning, in oil and gas wells, 88–93
  CH completion wells, formation damage removal in, 92–93
  OH completion wells, oil-based fluid filter cake removal in, 89–92
 during oil-based drilling fluids to water-based fluid displacement, 85–88
Wettability, of microemulsions, 76–77
Winsor phase behavior, 67–70, 67f–68f, 81
Wound debridement, microbial cleaning for, 154

X

X-ray diffraction, 82

Z

Zisman plot, 165–166, 166f
 of aqueous SDS solutions on an aluminum surface, 165f

Color Plates

FIGURE 1.3 Physical appearance of ionic liquids. On the left is methyl-tri-*n*-butylammonium dioctyl sulfosuccinate with a melting point around 313 K. On the right is 1-butyl-3-methyl imidazolium (diethylene glycol monomethyl ether) sulfate which is liquid at room temperature [69].

FIGURE 1.4 A room temperature ionic liquid compared with common table salt [47].

FIGURE 1.5 Metal-based ionic liquids exhibit a wide range of colors. The liquids are from left to right: copper-based compound, cobalt-based compound, manganese-based compound, iron-based compound, nickel-based compound, and vanadium-based compound [84]. *Source: Courtesy of Sandia National Laboratories, Albuquerque, NM.*

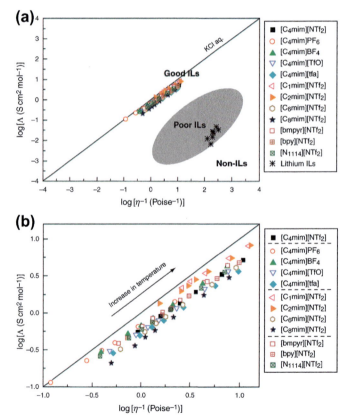

FIGURE 1.12 Walden plot of log (molar conductivity, Λ) against log (reciprocal viscosity η^{-1}) for ionic liquids. The upper figure (a) includes a classification of good and poor ionic liquids, as well as nonionic liquids [14]. The lower figure (b) is a close-up view of the region occupied by typical aprotic ionic liquids. The solid line indicates the ideal line for a completely dissociated strong electrolyte aqueous solution (KCl aq.) [221].

(a) Imidazolium salt

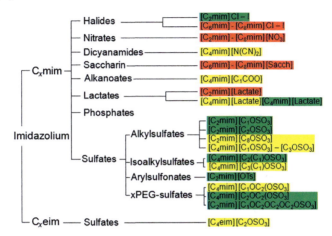

(b) Ammonium salt

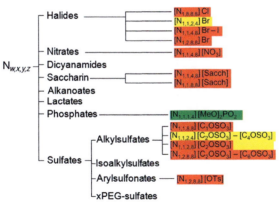

FIGURE 1.15 Proposed dendrograms to represent the toxicities of selected (a) imidazolium and (b) ammonium ionic liquids [276,282].

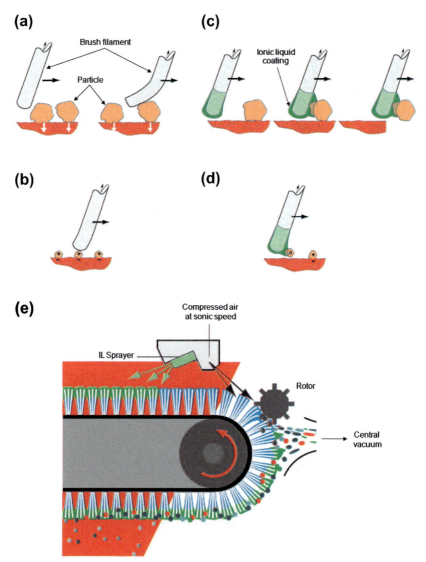

FIGURE 1.17 The removal of contaminant particles by brush cleaning (a and b) is much more efficient if the brush filaments are coated with a conducting ionic liquid film, applied by a spray from a fine nozzle (c, d and e). The adhering particles are removed by a rotor at the end of the process [334,335]. *Source: Courtesy of IoLiTec GmbH and Wandres Micro-Cleaning, Germany.*

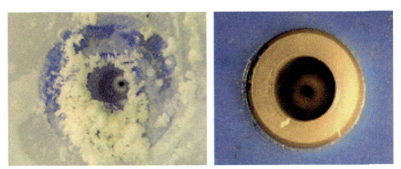

FIGURE 1.18 A spray nozzle for aqueous solutions of sodium chloride (left) and a hydrophilic ionic liquid (right), each after 10 h of operation [334]. *Source: Courtesy of IoLiTec GmbH, Germany.*

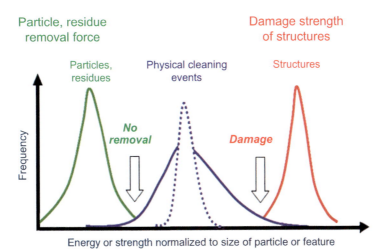

FIGURE 3.1 Cleaning process window depicting particle adhesion force distribution (left) compared with structural integrity force (right) with optimized cleaning forces located in between.

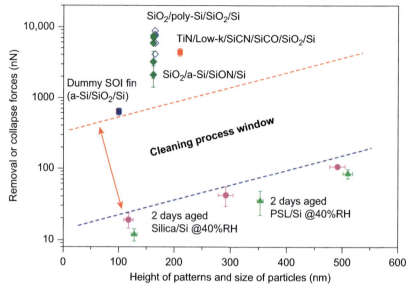

FIGURE 3.5 Cleaning process window showing less energy required to remove particles vs damage structures. Figure provided courtesy of Tae-Gon Kim and used with permission.

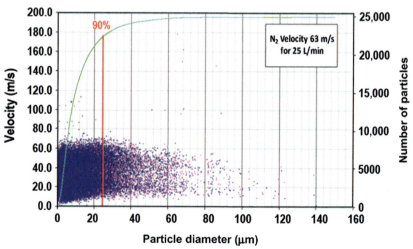

FIGURE 3.13 Droplet size and velocity distribution from atomized liquid spray with gas flow rate of 25 L/min. *Reproduced by permission of ECS - The Electrochemical Society.*

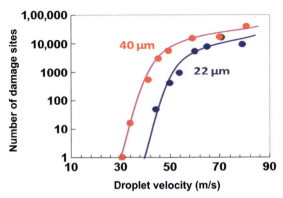

FIGURE 3.19 Damage to photoresist structures for two droplet sizes at varying velocity.

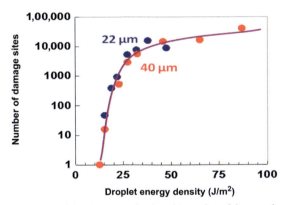

FIGURE 3.20 Correlation of droplet energy density with number of damage sites.

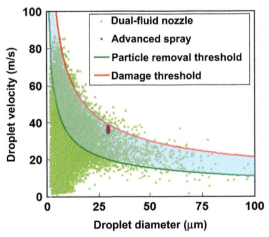

FIGURE 3.25 Threshold curves for particle removal and pattern damage generation with dual-fluid Nanospray2 and NanosprayÅ nozzles.

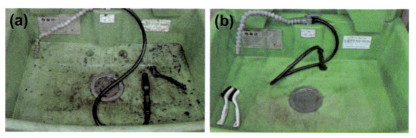

FIGURE 4.4 Parts cleaner sink (a) prior to cleaning, and (b) after cleaning [30]. *Source: Courtesy of J. Walter Co. Ltd.*

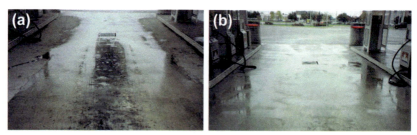

FIGURE 4.5 Photos of a truck fueling bay before (a) and after (b) microbial cleaning [36]. *Source: Courtesy Worldware Enterprises, Canada.*

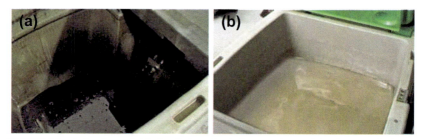

FIGURE 4.6 A cleaning tank (a) after drainage but before cleaning, and (b) after microbial treatment [34]. *Source: Courtesy of J. Walter Co. Ltd.*

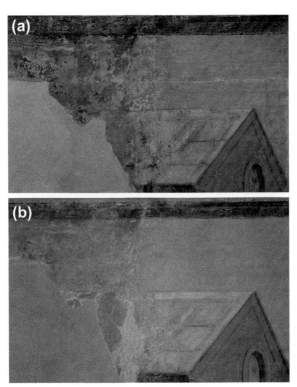

FIGURE 4.8 Effect of biocleaning with *Pseudomonas stutzeri* bacterial strain on the *Stories of the Holy Fathers* fresco before (a) and after (b) treatment [97].

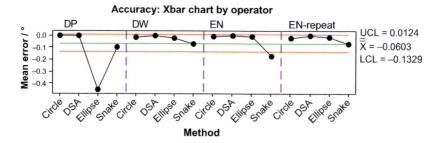
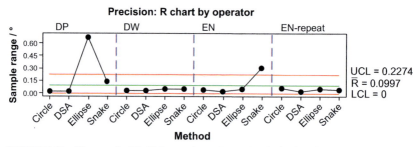

FIGURE 5.15 The use of a 3D SRO to evaluate the results obtained by four operators (DP, DW, EN, and EN-repeat) using four contact angle measurement procedures (Circle, DSA, Ellipse, and Snake). The horizontal reference lines in each chart are the 95% upper confidence limit (UCL), the mean of all measurements in the study (X), the 95% lower confidence limit (LCL), and the mean of all ranges in the study (R).

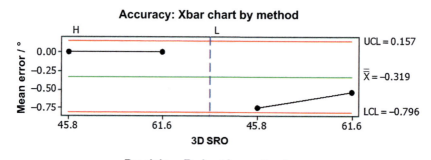
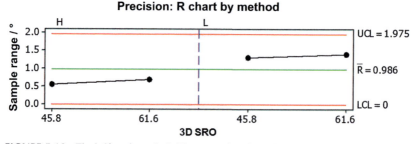

FIGURE 5.16 The half-angle method (H) was used to determine the accuracy and precision of the Langmuir method (L) against two 3D SROs (45.8° and 61.6°). The horizontal reference lines in each chart are the upper 95% confidence limit (UCL), the mean of all measurements in the study (X), the lower 95% confidence limit (LCL), and the mean of all ranges in the study (R).